T0141443

FLORA OF TROPICAL EAST AFRICA

———

ASPLENIACEAE

HENK BEENTJE[1]

Terrestrial or epiphytic ferns; rhizomes erect or creeping, with clathrate rhizome scales, usually dark-coloured. Fronds tufted or widely spaced, simple or variously pinnately dissected; stipes not articulated to the rhizome, with 2 vascular strands at base, uniting higher up to form a 4-armed X-shaped strand; veins free or anastomosing marginally. Sori borne on the costal side of a vein, elliptic to linear, with a cup-shaped to linear indusium, or indusium obsolete; sporangia with long stalks; spores monolete, with a perispore.

Two genera, about 600 species.

1. Lower surface of pinnae glabrous or with scattered scales; indusium present . . . . 1. **Asplenium**
 Lower surface of pinnae covered with dense imbricate scales; indusium obscure or absent . . . . 2. **Ceterach**

GLOSSARY OF SPECIALIST TERMS USED

acroscopic – on the side towards the apex
basiscopic – on the side towards the base
clathrate – latticed appearance
costa – midrib of a pinna
costule – midrib of a pinnule, or of the lobe of a pinna
decrescent – decreasing (often said of pinnae size towards the apex)
dimidiate – of pinnae or pinnules where the midrib forms the lower margin for quite some distance
flabellate – fan-shaped
frond – leaf of a fern, including the stalk (stipe), rachis and lamina or pinnae
gemma (plural gemmae) – bud arising on the frond, from which new plants can grow
indusium (plural indusia) – thin flap of tissue covering the young sorus
lithophyte – plant growing on rock

pinna – primary division of a compound leaf
pinnatifid – shallowly lobed
pinnatipartite – lobed to about halfway
pinnatisect – lobed almost to midrib
pinnule – the secondary or tertiary division of a compound leaf
proliferous – bearing gemmae, or plantlets
sorus (plural sori) – fertile body on a fern, composed of several spore-producing sporangia
sporangium (plural sporangia) – the spore case
spore – asexual reproductive body, usually powder-like
sterile – in ferns, without sporangia
stipe – the stalk of a frond or leaf
tufted – of fronds, with the stipes in dense clumps (as opposed to spaced apart)
ventral – lower surface

[1] This is by no means a definitive treatment, and several groups of taxa remain unresolved. However, with the key and descriptions I believe it will be possible to name most specimens; that is all I set out to do here.
 I would like to think Prof. Dr. Brigitte Zimmer for correcting some of my mistakes, as well as for the access to the Berlin herbarium with its important types. Dr. Bernard Verdcourt kindly read through the manuscript and improved it through various suggestions. Dr. Peter Chaerle kindly showed me his thesis on high-altitude *Asplenium* and corrected several mistakes.

1

1. ASPLENIUM

L., Sp. Pl. 2: 1078 (1753); Gen. Pl. ed. 5: 485 (1754)

Loxoscaphe Moore in Journ. Bot. [Hook.] 5: 227 (1853)

Rhizome erect or creeping, with clathrate scales, vascular cylinders with large overlapping leaf gaps (dictyostelic). Stipe matt or glossy, black, chestnut brown or greyish-green, glabrous or with hairs, or with clathrate scales. Frond simple to 4-pinnatifid, glabrous, pubescent or with scattered scales, veins pinnate or flabellate, free. Sori usually elongate, but about as long as broad in some species, borne on the costal side of a vein; indusium narrow.

Over 600 species, cosmopolitan.

Both Christensen in Pteridophyta of Madagascar pp. 85–86 and Ballard in the protologue of *A. paucijugum* point out the astonishing degree of variation occurring in many of the species of *Asplenium*.

Key to the species

Note: pinnae measurements are those of largest pinnae (usually near but not at the base of the lamina); sterile pinnae have not been taken into account as they may represent juvenile plants!

1. Fronds simple 2
   Fronds pinnatifid to quadripinnate 7
2. Lamina > 25 cm long 3
   Lamina 4–20 cm long 5
3. Lamina 70–200 × 10–22 cm; marginal vein
     present . . . .   1. *A. nidus* (p. 8)
   Lamina 25–100 × 2–8 cm; marginal vein absent 4
4. Rhizome scales to 13 × 2.7 mm, margins with
     protuberances . . . .   2. *A. holstii* (p. 10)
   Rhizome scales to 11 × 1.5 mm, with entire
     margin . . . .   3. *A. africanum* (p. 10)
5. Stipe 1–2 cm long . . . .   4. *A. peteri* (p. 11)
   Stipe 5–30 cm long 6
6. Fronds often proliferous at apex; rhizome to
     6 mm thick . . . .   5. *A. paucijugum* (p. 11)
   Fronds not proliferous; rhizome to 20 mm thick   6. *A. pocsii* (p. 12)
7. Fronds 1-pinnate, the margins of the pinnae
     entire or crenate, but not lobed; e.g. fig. 2, p. 14 8
   Fronds with at least the lowermost pinnae 2-
     pinnatifid, or pinnae lobed or more dissected 54

1-PINNATE
8. Fronds proliferous 9
   Fronds not proliferous (check all fronds!) 24

1-PINNATE PROLIFEROUS
9. Gemmae halfway up the terminal pinna . . . . . .   7. *A. angolense* (p. 13)
   Gemmae either just below the terminal pinna or
     at each pinna apex 10
   [Gemmae halfway up most pinnae – see note sub 16. *A. elliottii*]

10. Pinnae in 1–5 pairs, the largest > 15 cm long 11
    Pinnae in > 6 pairs (or if few, then < 10 cm long) 13
11. Pinna apex emarginate and proliferous – at
    (almost) every apex . . . .   8. *A. emarginatum* (p. 13)
    Pinna apex acuminate or attenuate 12
12. Pinnae in 1–2 pairs; sori 7–29 mm long . . . .   5. *A. paucijugum* (p. 11)
    Pinnae in 4–5 pairs; sori 2.5–10 mm long . . . . .   (unresolved taxa)
                         L2775 ined. (p. 67)
13. Fronds spaced . . . .   9. *A. gemmascens* (p. 15)
    Fronds tufted (rarely spaced in 18. *macrophlebium*) 14
14. Rachis lengthened and ending in gemma/bud,
    looping and rooting to form large colonies . .   10. *A. sandersonii* (p. 15)
    Gemma present at base of terminal segment 15
15. Largest pinnae < 2 cm long; pinnae many,
    18–45 pairs 16
    Largest pinnae > 2 cm long 17
16. Sori solitary (–2); basal pinnae reduced . . . . . .   11. *A. monanthes* (p. 16)
    Sori 2–7; basal pinnae not very reduced . . . . . .   12. *A. normale* (p. 17)
17. Terminal segment ± similar to lateral pinnae,
    > 2 cm wide; sori 5–30 mm long 18
    Terminal segments gradually decrescent, not
    similar to lateral pinnae, or similar and < 2 cm
    wide; sori < 10 mm long 19
18. Rhizome scales 9–14 × 2–3 mm . . . .   13. *A. gemmiferum* (p. 18)
    Rhizome scales to 3.5 × 0.5 mm . . . .   21. *A. warneckei* (p. 26)
19. Rachis with small stalked glands; $K\,4$ . . . .   14. *A. adamsii* (p. 18)
    Rachis glabrous or with scales 20
20. Largest pinnae to 16 cm long; rhizome scales
    > 4 mm long 21
    Largest pinnae to 6(–8) cm long; rhizome scales
    up to 4 mm long 22
21. Rhizome scales linear-lanceolate, to 18 × 2 mm;
    pinna margin incised into bilobed or crenate
    lobes . . . .   15. *A. boltonii* (p. 20)
    Rhizome scales ovate, 4–10 × 2.5–4 mm; pinna
    margin minutely crenate-serrate . . . .   16. *A. elliottii* (p. 21)
22. Sori 2–4 mm long; pinnae to 4 × 1.2 cm; rhizome
    scales with marginal protuberances; $U\,2$ . . . .   17. *A. barteri* (p. 22)
    Sori 2–10 mm long; pinnae to 6 × 2.3 cm;
    rhizome scales entire 23
23. Veins unbranched – except for the basal ones . .   18. *A. macrophlebium* (p. 22)
    All veins forked . . . .   19. *A. christii* (p. 23)

1-PINNATE NON-PROLIFEROUS
24. Fronds spaced, from a creeping rhizome 25
    Fronds tufted, rhizome erect or shortly creeping 30

1-PINNATE NON-PROLIFEROUS; SPACED FRONDS
25. Pinnae to 4.8 cm wide, in 4–15 pairs; sori usually
    > 10 mm long 26
    Pinnae to 2.3(–3) cm wide, in 12–20 pairs; sori
    < 9 mm long 27
26. Pinnae wedge-shaped, ± 3-lobed at apex with
    long central lobe . . . .   20. *A. megalura* (p. 24)
    Pinnae asymmetrically and rather broadly
    ovate-rhomboid . . . .   21. *A. warneckei* (p. 26)

60. Pinnae with basiscopic margin dimidiate for
    > 50%; pinnae < 2 cm long . . . .   42. *A. dregeanum* (p. 43)
    Pinnae with basiscopic margin dimidiate for
    < 30%; pinnae > 3 cm long . . . .   43. *A. preussii* (p. 44)

NON-PROLIFEROUS
61. Lamina 3–4-pinnate or 4–5-pinnatifid/-sect 62
    Lamina 2-pinnatifid, 2-pinnate or up to 3-pinnate 65
62. Rhizome scales up to 13 mm; sori marginal and
    cup-shaped, 1–1.5 mm long . . . .   47. *A. hypomelas* (p. 48)
    Rhizome scales up to 6 mm; sori linear, 1.5–6 mm
    long 63
63. Lamina narrowly elliptic in outline, basal
    pinnae (slightly) reduced; pinnae to 9 × 3 cm;
    sori marginal . . . .   46. *A. S618* ined. (p. 48)
    Lamina ovate-triangular in outline, basal pinnae
    largest; pinnae to 17 × 10 cm 64
64. Ultimate lamina segments narrow and slightly
    obovate, with apical teeth to 2 mm long; sori
    1.5–6 mm long . . . .   45. *A. linckii* (p. 46)
    Ultimate segments broad and obtriangular, with
    minute crenations; sori ± 1 mm long . . . .   (see Note under
                                                   A. abyssinicum) (p. 63)
65. Rhizome creeping; fronds well-spaced 66
    Rhizome erect or short-creeping; fronds tufted
    or short-spaced 74
66. Sori parallel to pinna midrib; pinnae in 14–35
    pairs; stipe 15–77 cm long . . . .   23. *A. friesiorum* (p. 27)
    Sori in lobes or pinnules at angle to pinna
    midrib; stipe 4–55 cm 67
67. Stipe and rachis with stalked glands . . . .   48. *A. actiniopteroides* (p. 49)
    Stalked glands absent (very rarely present in
    A. praegracile) 68
68. Pinnae less than 3 × 1.2 cm; stipe to 10 cm long 69
    Pinnae nearly always (much) longer (though
    may be small in *uhligii*); stipe 4–55 cm 70
69. Pinnae up to 2.5 × 1.3 cm; stipe to 6 cm; lamina
    2-pinnate to 3-pinnatifid; sori 3–5 mm long . .   49. *A. goetzei* (p. 50)
    Pinnae to 3 × 1.2 cm; stipe 6–10 cm; lamina
    pinnate with deeply lobed pinnae; sori 1–4 mm   54. *A. lividum* (p. 53)
70. Fronds 2-pinnatifid to 2-pinnatisect, the basal
    pair of pinnules sometimes divided to 3
    orders (note: some *A. lividum* might also key
    here – they are from 1000–2100 m and
    generally have hardly spaced fronds) 71
    Fronds 3-pinnatifid to 3-pinnatisect 73
71. Rhizome scales 5–9 × 1–2 mm; altitude
    (2400–)2600–4200 m . . . .   50. *A. uhligii* (p. 50)
    Rhizome scales 1–5 × 0.3–0.5 mm 72
72. Rhizome scales 2–5 × 0.3 mm; altitude
    30–500(–1200) m . . . .   59. *A. buettneri*
                                                 var. *hildebrandtii* (p. 60)
    Rhizome scales 1–4 × 0.5 mm; altitude
    2300–3000 m . . . .   52. *A. mildbraedii* (p. 52)

73. Rhizome scales to 0.4 mm wide, with paler
    margin . . . .   51. *A. praegracile* (p. 51)
    Rhizome scales to 1.5 mm wide, margin same
    colour as rest of scale . . . .   53. *A. volkensii* (p. 52)

NON-PROLIFEROUS; TUFTED FRONDS
74. Pinnae with narrow linear lobes (< 1.5 mm
    wide), though the basal one(s) may be lobed
    themselves 75
    Pinnae hardly lobed, or lobes wider and not
    linear 80
75. Rhizome scales < 3 mm long 76
    Rhizome scales up to 12 mm long; sori 1 per
    lobe 77
76. Sori 1–12 per lobe, and also in wings along
    costa . . . .   54. *A. lividum* (p. 53)
    Sori 1–2(–3) per lobe . . . .   27. *A. formosum* (p. 31)
77. Sori near apex of lobes, to 2 × 2 mm, forming a
    marginal 'pocket' . . . .   55. *A. theciferum* (p. 55)
    Sori halfway up the lobe, usually longer than
    wide 78
78. Most pinnules lobed . . . .   46. *A. S618* ined. (p. 48)
    Only basal pinnules lobed 79
79. Rhizome scales dark brown with or without a pale
    margin, narrowly triangular, 4–9 × 0.8–1.2
    (–2.2) mm, ending in a hair-tip; stipe 3–20 cm
    long; lamina 7–40 × 3–10(–12) cm; pinnae in
    13–28 pairs, the largest 3–7 × 0.7–1.3(–1.7) cm;
    widespread, 750–2100(–2300) m . . . .   56. *A. rutifolium* (p. 56)
    Rhizome scales pale to mid-brown, ovate, to 12 ×
    2–3 mm; stipe 15–48 cm long; lamina 32–80 ×
    13–26 cm; pinnae in 23–44 pairs, the largest
    (6–)8–13(–16) × 1–2.2 cm; widespread,
    1900–3100(–3650) m . . . .   57. *A. loxoscaphoides* (p. 57)
    (*A. sertularioides* is ± intermediate between these two; **T** 2, 3350–3500 m)
80. Sori along the costa, sometimes also in ultimate
    segments 81
    Sori in ultimate segments only 82
81. Sori in ultimate segments as well as in wings
    along the costa . . . .   54. *A. lividum* (p. 53)
    Sori along the costa only, not in the lobes or
    only in the basal lobe . . . .   22. *A. pellucidum* (p. 26)
82. Sori 1.5–14 mm long (sori may look continuous
    in *A. aethiopicum*!) 83
    Sori 1–5 mm long 87
83. Lamina 2-pinnatifid 84
    Lamina 2-pinnate to 3-pinnatifid 86
84. Pinnae in 8–11 pairs, 3-lobed with prominent
    central lobe; rhizome scales 3–8 mm long . . .   32. *A. stuhlmannii* (p. 34)
    Pinnae without prominent central lobe 85
85. Pinnae in 7–25 pairs, deeply pinnatifid with > 3
    pinnules; rhizome scales 3–7 mm long . . . . . .   58. *A. aethiopicum* (p. 58)
    Pinnae 14–24 pairs, lanceolate with regularly
    lobed margin; rhizome scales 7–11 mm long   29. *A. smedsii* (p. 32)

Dimorphic fronds (with different shapes and sizes for fertile and sterile fronds) occur in *A. christii.*

1. **Asplenium nidus** *L.*, Sp. Pl. 2: 1079 (1753); Johns, Pterid. trop. East Africa checklist: 66 (1991). Type: *Osbeck* 49, Herb. Linn. No. 1250.6 (LINN, lecto.), designated by Holttum in Gard. Bull. Singapore 27: 147 (1974)

Epiphyte or occasional lithophyte, forming large clumps up to 2 m diameter; rhizome erect, thick (at least to 2.5 cm but probably more), bearing many persistent rootlets and with golden brown narrowly triangular rhizome scales to 25 × 3 mm, the margins with protuberances, apex attenuate into a hair-tip. Fronds tufted. Stipe blackish, almost absent or to 5 cm long. Lamina very shiny with black midrib (fide *Verdcourt*), coriaceous, narrowly elliptic, 70–200 × (10.5–)15–22 cm, simple, apex acuminate; costa raised above, almost flat beneath; veins forked from near base, dense and at 30–45° to midrib, united just before margin with a marginal vein; veins and costa scaly when young, glabrescent. Sori along the veins in the distal half of the lamina, from near midrib/costa to ¼ to ½ of the width of the lamina, linear, (5–)17–58 mm long; indusium membranous, entire, to 0.5 mm wide. Fig. 1: 1–2, p. 9.

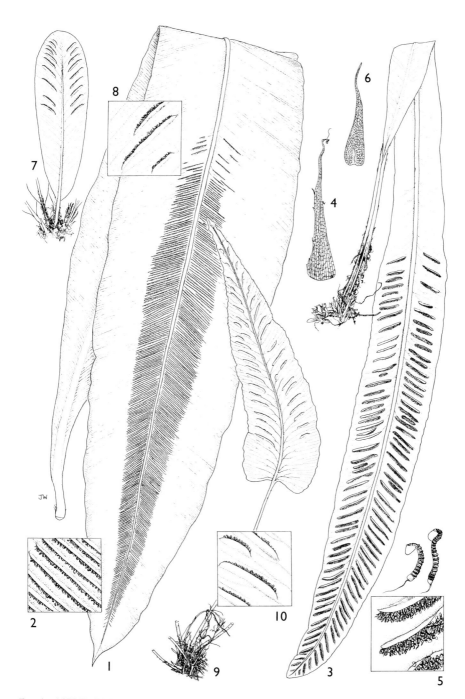

FIG. 1. *ASPLENIUM NIDUS* — **1**, habit × ¹/₄; **2**, sori × 2. *ASPLENIUM HOLSTII* — **3**, habit × ¹/₂; **4**, rhizome scale × 3; **5**, sori × 2¹/₂. *ASPLENIUM AFRICANUM* — **6**, rhizome scale × 3. *ASPLENIUM PETERII* — **7**, habit × ¹/₂; **8**, sori × 2. *ASPLENIUM PAUCIJUGUM* — **9**, habit × ¹/₂; **10**, sori × 2. 1, from *Richards* 18177; 2, from *Verdcourt* 69; 3, 5, from *Faden* 71/26; 4, from *Peter* 24238; 6, from *Faden* 70/39; 7, 8, from *Luke et al.* 5368; 9, 10, from *Balslev* 282. Drawn by Juliet Williamson.

KENYA. Kilifi District: Kaya Jibana, Aug. 1996, *Luke et al.* 4529!; Kwale District: Shimba Hills, Kivumoni, March 1991, *Luke & Robertson* 2693! & idem, Longomwagandi, June 1996, *Luke et al.* 4512!

TANZANIA. Lushoto District: Usambara Mts. Amani, Sigi R., Aug. 1963, *Ali Omari in Richards* 18177! & Mtai Forest Reserve, Oct. 1999, *Kayombo et al.* 2935!; Morogoro District: Nguru Mts, NW slope of Mkobwe, March 1953, *Drummond & Hemsley* 1887!; Zanzibar, Jozani forest, Dec. 1929, *Vaughan* 998!

DISTR. **K** 7; **T** 3, 6; **Z, P**; Madagascar, Indian Ocean islands; India, China, Japan, SE Asia, N Australia, Micronesia, Pacific islands

HAB. Moist forest, on Pemba also in 'closed bush', may be locally common; low to medium-high epiphyte, occasionally on rocks; 0–1200 m

CONSERVATION NOTES. Widespread; least concern (LC)

2. **Asplenium holstii** *Hieron.* in E.J. 46: 348, t. 53/h (1911); Schelpe, F.Z. Pteridophyta: 170, t. 53h (1970); Johns, Pterid. trop. East Africa checklist: 63 (1991). Type: Tanzania, Usambara, Bumbu-Fustii near Masheua, *Holst* 8726 (B!, syn. – or lecto., chosen by R.Viane – on detslip at B, dated Sept. 1985); Shagaia near Mbaramu, *Holst* 3701 (B!, syn.)

Epiphyte, lithophyte or rarely terrestrial; rhizome erect or shortly ascending, 5–20 mm diameter and 7 cm long, with narrowly triangular clathrate (dark cell walls, translucent lumen) rhizome scales up to 13 × 2.7 mm, wider and almost ovate at base, margin irregularly and sparsely hair-lobed, apex very acuminate and ending in a hair. Fronds tufted, rarely shortly spaced, erect or pendulous. Stipe pale matt grey-brown, gradually attenuate from lamina but sometimes up to 20 cm long and to 4 mm diameter, at the very base with scales similar to rhizome, elsewhere at first clothed with scales similar to those of rhizome but smaller, to 6 × 0.5 mm, soon becoming subglabrous. Lamina dark glossy green above, paler beneath, fleshy, elliptic to linear-oblong, 30–100 × (2–)2.5–7 cm, simple, base attenuate, margin obscurely undulate or subentire to undulate-crenulate near apex, apex acute to acuminate or occasionally rounded or caudate, glabrous or occasionally with fimbriate-ciliate scales to 4 × 0.25 mm with hair-like apex; costa ± flat; lateral veins visible, furcate, not reaching margin. Sori along the veins, usually extending from near the costa to $\frac{1}{2}$–$\frac{3}{4}$ of the distance to the margin, linear, 5–33 mm long; indusium membranous, entire, 0.5–1.4 mm wide. Fig. 1: 3–5, p. 9.

KENYA. Teita District: Taita Hills, Mbololo, Ndaru ridge, July 1969, *Faden et al.* 69/834! & Mbololo, May 1985, *Taita Hills Expedition* 418!; Kasigau, Rukanga route, Nov. 1994, *Luke & Luke* 4145!

TANZANIA. Lushoto District: Baga Forest Reserve, May 1987, *Kisena* 535!; Morogoro District: Nguru Mts, Kwegoba Mt, June 1978, *Thulin & Mhoro* 3095!; Iringa District: Udzungwa Mountain National Park, Sonjo–Mwanihana route, Nov. 1997, *Luke & Luke* 5020!

DISTR. **K** 6, 7; **T** 2, 3, 6, 7; Malawi, Mozambique, Zimbabwe

HAB. Moist forest, often near streams; low epiphyte, on moist rocks, or rarely terrestrial; 400–1950 m

CONSERVATION NOTES. Widespread; least concern (LC)

SYN. *Asplenium africanum* Desv. var. *holstii* (Hieron.) Tardieu in Mém. Inst. Fr. Afr. Noire, 28: 172 (1953)

NOTE. Near *A. africanum* Desv.; only differs in rhizome scales which are slightly wider, with a margin that has protuberances; and geographically.

Bob Johns has made a special study of this group and feels that there are some twelve taxa, distinguishable by spore characters.

3. **Asplenium africanum** *Desv.* in Mag. Ges. Naturf. Freunde Berlin 5: 322 (1811); Tardieu, Mém. I.F.A.N. 28: 171, t. 32/1–2 (1953); Alston, Ferns W.T.A.: 55 (1959); Tardieu, Fl. Cameroun Pterid.: 178 (1964); Johns, Pterid. trop. East Africa checklist: 61 (1991); Faden in U.K.W.F. ed. 2: 27 (1994). Type: Benin [Oware kingdom], *Palisot de Beauvois* s.n. (P, holo., not found)

Epiphyte or lithophyte; rhizome erect or shortly creeping (fide *Faden*), short, fleshy, with lanceolate clathrate (dark brown cell walls, translucent lumen) rhizome scales to 11 × 1.5 mm, margin entire to sinuate or crenate (not crenate in our specimens), apex attenuate and ending in a hair-tip. Fronds tufted, drooping. Stipe thick, canaliculate, gradually attenuate from lamina but sometimes seemingly to 25 cm long, with a few scales at the very base. Lamina slightly fleshy, lanceolate or strap-shaped, 25–100 × (2.2–)3–8 cm, entire, base attenuate, apex attenuate, glabrous; costa clear, flattened; veins many, free, forking, 6–7 per cm, at angle of 45–70° with costa. Sori many along the veins, linear, long ones alternating with occasional short ones, 9–30 mm long, reaching neither costa nor margin; indusium membranous, entire, pale, to 0.7 mm wide. Fig. 1: 6, p. 9.

UGANDA. Bunyoro District: Budongo Forest Reserve near Sonso R., Sep. 1995, *Poulsen et al.* 940!; Toro District: Kibale forest, July 1938, *A.S. Thomas* 2284!; Mengo District: Entebbe, Jan. 1938, *Chandler* 2136! & Sezibwa [Sezibura] falls, Sep. 1961, *Rose* 203!
KENYA. North Kavirondo District: Kakamega forest, Kibiri block, Yala R., Jan. 1970, *Faden et al.* 70/39!
TANZANIA. Bukoba District: Minziro Forest Reserve, Bulembe path, Sep. 2000, *Festo & Bayona* 766!
DISTR. **U** 2, 4; **K** 5; **T** 1; West Africa from Guinea to Cameroon and Congo-Kinshasa, south to Angola
HAB. Moist forest, often near water, (low) epiphyte or occasionally on rocks; 1000–1550 m
CONSERVATION NOTES. Widespread; least concern (LC)

SYN. *Asplenium sp. A* of Johns, Pterid. trop. East Africa checklist: 68 (1991)

NOTE. *Chandler* 2445a from **U** 4, Sezibwa Falls, has a very small fronds but has sporulating sori; the lamina on fertile fronds can be as small as 12 × 0.8 cm. Rhizome scales and general look indicate that this is an extreme form.
   Dummer 2728 from Uganda has some leaves with narrowly triangular outgrowths to 25 mm long on the margin; again, a form rather than a new taxon.
   Johns split off *Faden* 70/39 as species A of his checklist, without indicating any differences. I believe it is the same taxon.

4. **Asplenium peteri** *Becherer* in Ber. Schweiz. Bot. Ges. 38: 180 (1929); Johns, Pterid. trop. East Africa checklist: 66 (1991) Type: Tanzania, Tanga District: Mlinga Mts, Magila to Magrotto, *Peter* 39946 (B!, holo.)

Low terrestrial (type) or epiphyte (*Balslev*); rhizome erect, with dark brown, triangular scales 1–1.3 × 0.4 mm, attenuate, the margins entire. Fronds tufted, 10–13 cm long. Stipe 1–2 cm long, glabrous or nearly so. Lamina herbaceous, green, lanceolate, 9–11 × 2–2.6 cm, simple, costa glabrous, white, base long-attenuate, margin crenate-serrate, apex rounded or obtuse, here often proliferous. Sori many, at an oblique angle, reaching neither costa nor margin, linear, 3.5–8 mm long; indusium membranous, to 0.7 mm wide, entire. Fig. 1: 7–8, p. 9.

KENYA. Teita District: Kasigau mountain, Rukanga route, June 1998, *Luke et al.* 5368!
TANZANIA. Bukoba District: Munene Forest Reserve, Kyaka road, Jan. 1975, *Balslev* 530!; Amani, without date, *Grote* 8585!
DISTR. **K** 7; **T** 1, 3; not known elsewhere
HAB. Mist or swamp forest; 1170–1480 m – though the type must have been from a lower (but unknown) altitude
CONSERVATION NOTES. Though with a large extent of occurrence, this is only known from three recent localities; here assessed as Vulnerable (VU-D2).

SYN. *Asplenium lanceolatum* Peter, F.D.O.-A.: 67, 72 (1929) & Descr.: 5, t. 5.3 (1929), *non* Huds. 1762 *nec* Forssk. 1775

5. **Asplenium paucijugum** *Ballard* in Hook. Icon. Pl. ser. 5. 3: t. 3287 (1935); Johns, Pterid. trop. East Africa checklist: 66 (1991); Schippers in Fern Gaz. 14, 6: 202 (1993). Type: Tanzania, Lushoto District: Gonja, *Holst* 4246 (K!, holo.)

Terrestrial, low epiphyte or on rocks; rhizome creeping or erect, 3–6 mm diameter, with pale brown or dark brown ovate acuminate clathrate deciduous scales 2–4 mm long, the margins of the scales entire or sometimes (on same plant) with a few narrow teeth. Fronds shortly spaced or tufted, erect, 20–60 cm long, entire or more often 1-pinnate, often proliferous at or near apex, occasionally proliferous at several pinnae apices. Stipe pale brown or green, once described as black with pale lines, 5–44 cm long, with scales similar to those of rhizome but soon glabrescent. Lamina dark green, slightly coriaceous, simple or imparipinnate with 1–2 pairs of (sub-)sessile subopposite pinnae (rarely one of the pinnae with a single sub-pinna), the pinnae or lamina oblong-elliptic to lanceolate, 4–22 × 1.5–6.5 cm, the terminal pinna slightly larger than the laterals, base slightly cordate to cuneate, margin sinuate to crenate-serrate, apex acuminate with acumen up to 2 cm long, glabrous; veins forked. Rachis similar to stipe. Sori many, evenly spaced and closely parallel along the oblique veins, reaching neither costa nor margin, 7–29 mm long; indusium membranous, linear, ± 0.3 mm wide, entire or with minutely fimbriate margin. Fig. 1: 9–10, p. 9.

UGANDA. Toro District: Kibale National Park, Dura R., June 1997, *Poulsen & Nkuutu* 1272!; Mengo District: Sesse Islands, 1904, *Dawe* 72! & Sezibwa [Ssezzibwa] Falls 10 km W of Lugazi, Sep. 1969, *Faden & Evans* 69/978!
KENYA. Teita District: Taita Hills, Mbololo, May 1985, *NMK Taita Hills Expedition* 366! & idem, July 1998, *Bytebier* 1179!
TANZANIA. Lushoto District: Bomole near Amani lake, Feb. 1950, *Verdcourt* 91!; Morogoro District: Kanga Mts, Feb. 1970, *Pócs* 6139D!; Iringa District: Udzungwa Mountains, Ndunduru Forest Reserve, Oct. 2000, *Luke et al.* 7078!
DISTR. **U** 2, 4; **K** 7; **T** 1?, 3, 6, 7; West Africa from Guinea to Ghana, Sao Tomé; Madagascar
HAB. Moist forest, often along streams and falls; may be locally common; 600–1700 m
CONSERVATION NOTES. Widespread; least concern (LC)

SYN. *Asplenium variabile* Hook. var. *paucijugum* (Ballard) Alston in Bol. Soc. Brot. 30: 7 (1956) & Ferns W.T.A.: 56 (1959)

NOTE. Ballard in his protologue states that this taxon is "not so homogeneous a species as one might wish; (...) certain differences in rhizome scales lead one to wonder whether some of the plants may possibly represent juvenile states of other well-known species".
The Ugandan specimens have longer rhizome scales than the more Eastern ones; I suspect there might be a Western taxon and an Eastern one.
A. paucijugum Ballard var. *simplex* Kunkel (1962) is a name I have been unable to trace. It might be a manuscript name.

6. **Asplenium pocsii** *Pic.Serm.* in Webbia 27: 432, t. 16 (1973); Johns, Pterid. trop. East Africa checklist: 66 (1991); Schippers in Fern Gaz. 14, 6: 203 (1993). Type: Tanzania, Morogoro District: Uluguru Mts, E side of Bondwa, *Faden* 70/631 (herb. Pic.Serm., holo.; EA!, K!, iso.)

Terrestrial or epiphyte; rhizome oblique or suberect, to 20 mm diameter, with rhizome scales brown, ovate, 3–4 × 1.2–1.4 mm, margin sometimes with a few teeth, acuminate. Fronds tufted, ± erect, not proliferous. Stipe 6–12 cm long, sparsely scaly. Lamina lanceolate, 7–20 × 1.8–3.8 cm, entire, margin rather obscurely crenate, apex attenuate into an obtuse tip; veins branched about halfway; with minute sparse scales beneath. Sori along the veins, almost reaching midrib and margin, linear, (5–)7–20 mm long; indusium entire, membranaceous, to 1.4 mm wide.

TANZANIA. Morogoro District: Uluguru Mts, E edge of Lukwangule Plateau, Dec. 1969, *Pócs et al.* 6082c; idem, Magari, date unclear, *Pócs et al.* 6297e; Ukaguru Mts, Mamiwa Forest Reserve, just below Mamiwa summit, Aug. 1972, *Mabberley & Salehe* 1488!
DISTR. **T** 6; not known elsewhere
HAB. Moist forest; 1650–2400 m

CONSERVATION NOTES. Here assessed as Endangered (EN-B2a, biii) due to the limited area of occupancy coupled to a threat to the forest habitat at lower altitudes.

NOTE. This taxon is in the "*variabile*" group but split off by Pichi Sermolli. Sadly, he did not indicate how *pocsii* differs from *paucijugum*; he merely describes differences with a taxon from Madagascar.

7. **Asplenium angolense** *Baker*, Syn. Fil.: 485 (1883); Johns, Pterid. trop. East Africa checklist: 61 (1991); Faden in U.K.W.F. ed. 2: 27 (1994). Type: Angola, *Welwitsch* 86 [in Kew specimen changed in pen to 96] (K!, iso.)

Terrestrial; rhizome erect or shortly creeping, to 5 mm diameter, with dark brown narrowly triangular ?entire acute rhizome scales to 1.5 mm long, with paler margins. Fronds tufted, to 30 cm long, proliferous in centre of terminal pinna. Stipe 9–25 cm long, with scattered scales similar to those of rhizome. Lamina ovate in outline, 8–13 × 7–10 cm, 1-pinnate, the lower pinnae hardly or not decrescent, the terminal pinna very similar to the lateral ones and the same size. Pinnae 2–3 pairs, (sub-)opposite, to 6 × 2 cm, shortly petiolate, narrowly ovate, slightly asymmetric with the acroscopic base parallel to the rachis, basiscopic base more cuneate, margin crenate, glabrous or nearly so. Rachis slightly winged, with some scattered scales. Sori many on veins, linear, to 10 mm long; indusium linear, entire, to 0.7 mm wide, membranous. Fig. 2: 1–2, p. 14.

KENYA. N Kavirondo District: Isiukhu R. S to SSW of Walikale on Kambiri–Vihiga road, Dec. 1969, *Faden & Rathbun* 69/2112! & Kakamega Forest, Kibiri block, S of Yale R., Jan. 1970, *Faden et al.* 70/1!
DISTR. **K** 5; Angola (see Note)
HAB. Dense shade in intermediate forest; 1600–1700 m
CONSERVATION NOTES. Data deficient, due to the uncertainty in taxonomy

NOTE. The correct name of this plant is uncertain. The position of the gemma in the centre of terminal pinna is constant and unique for EA.
   Two specimens seem to belong here, but lack the gemmae so key out to *A. inaequilaterale*. Uganda, **U** 2, Kigezi District: Bwindi National Park, Feb. 1995, *Poulsen et al.* 729! and Tanzania, **T** 3, Lushoto District: Shengena Forest Reserve, Gonja, Feb. 1988, *Kisena* 359!

8. **Asplenium emarginatum** *P.Beauv.*, Fl. Oware 2: 6, t. 61 (1807); Alston, Ferns W.T.A.: 56 (1959); Johns, Pterid. trop. East Africa checklist: 63 (1991). Type: [L'isle du Prince], *Palisot de Beauvois* s.n. (P, holo., not found)

Terrestrial; rhizome erect or short-creeping, to 10 mm diameter, with persistent ovate long-acuminate subentire clathrate scales to 7 mm long. Fronds tufted or sometimes shortly spaced, 35–90 cm long, proliferous at almost every pinna apex, the terminal lobe similar to the laterals but usually larger. Stipe pale brown, 10–50 cm long, sulcate, with subulate or capillary scales to 6 mm long. Lamina erect, (narrowly) ovate in outline, 1-pinnate, proliferous at pinnae apices. Pinnae in 1–5 pairs, ovate, 6–23 × 2.4–5.5 cm, sessile or very shortly petiolate, base unequally cuneate, margin crenate, apex emarginate with an abrupt and cuneate sinus, glabrous on both surfaces; veins forked. Rachis similar to stipe. Sori many and usually closely parallel on the ascending veins, almost reaching the costa but ending quite a way from the margin, brown and oblong, to 2.5 cm long; indusium whitish, membranous, entire, linear, persistent. Fig. 2: 3–5, p. 14.

UGANDA. Bunyoro District: Budongo Forest Reserve, Kanyo-Pabidi block, Feb. 1996, *Poulsen et al.* 1104!; Mengo District: Kajansi Forest, June 1937, *Chandler* 1652! & Sezibwa [Ssezzibwa] Falls 10 km W of Lugazi, Sep. 1969, *Faden & Evans* 69/981!
KENYA Kilifi District: 1 km NE of Pangani on Chonyi–Ribe road, May 1972, *Faden & Faden* 72/224!; Kwale District: Shimba Hills, Mwele Forest, Jan. 1989, *Luke* 1645! & Dzombo Hill, Feb. 1989, *Mrima–Dzombo Expedition* 294!

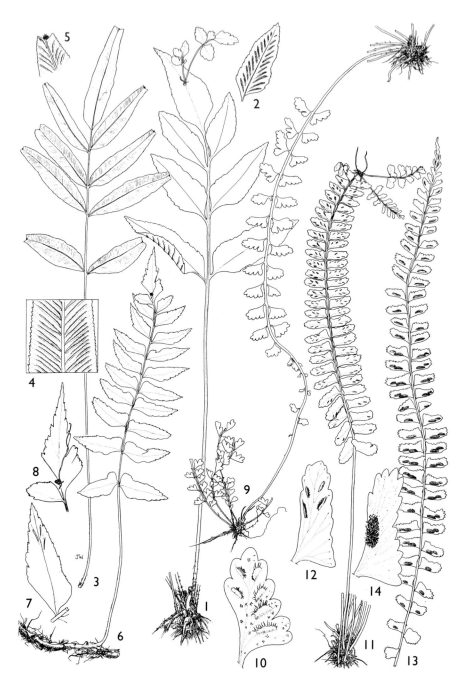

FIG. 2. *ASPLENIUM ANGOLENSE* — **1**, habit × ¹/₂; **2**, sori detail × ¹/₂. *ASPLENIUM EMARGINATUM* — **3**, habit × ¹/₈; **4**, sori × ¹/₂; **5**, gemma × ¹/₂. *ASPLENIUM GEMMASCENS* — **6**, habit × ¹/₄; **7**, sori × ¹/₂; **8**, gemma × ¹/₂. *ASPLENIUM SANDERSONII* — **9**, habit × ¹/₂; **10**, sori × 2. *ASPLENIUM MONANTHES* — **11**, habit × ¹/₂; **12**, sori detail × 2. *ASPLENIUM NORMALE* — **13**, habit × ¹/₂; **14**, sori × 2. 1, from *Faden et al.* 70/1; 2, from *Faden & Rathbun* 69/2112; 3, 4, 5, from *Faden & Evans* 69/981; 6, 7, 8, from *Faden* 69/1033; 9, 10, from *Faden & Cameron* 72/298; 11, 12, from *Horlyck & Joker* TZ 329; 13, 14, from *Tweedie* 1894. Drawn by Juliet Williamson.

TANZANIA. Lushoto District: Baga Forest Reserve, May 1987, *Kisena* 633!; Ulanga District: Magombera Forest Reserve, Feb. 1977, *Vollesen* MRC 4436!; Lindi District: Rondo Plateau, Rondo Forest Reserve, Feb. 1991, *Bidgood et al.* 1462!
DISTR. **U** 2, 4; **K** 7; **T** 3, 6–8; West Africa from Guinea to Sudan, and south to Angola
HAB. Moist forest, usually in deep shade; less often in semi-deciduous forest; sometimes locally common but more often rare; 40–1200(–1550) m
CONSERVATION NOTES. Widespread; least concern (LC) though nowhere common

NOTE. Two specimens from **T** 3: Lushoto District: Baga Forest Reserve, May 1987, *Kisena* 633! & 634! are similar in all respects – apart from not having any gemmae.
    Reichstein et al. 2939 from Kenya **K** 7: Teita District: Mbololo Hill, Nov. 1969, is similar but has attenuate apices to the pinnae; as most pinnae are proliferous, I assume this is (or is near) *A. emarginatum.*

9. **Asplenium gemmascens** *Alston* in Bol. Soc. Brot. 30: 10 (1956) & Ferns W.T.A.: 59 (1959); Tardieu, Fl. Cameroun Pterid.: 207, t. 33/1–3 (1964); Johns, Pterid. trop. East Africa checklist: 63 (1991). Type: Nigeria, Ogoja, Ikwette-Baleghete Pass, *Savory & Keay* FHI 25201 (BM!, holo.)

Epiphytic or terrestrial; rhizome long-creeping, to 7 mm diameter, with dark brown opaque attenuate rhizome scales (not apparent in East Africa). Fronds spaced, 60–80 cm long; stipe dark brown or matt grey, 15–30 cm long, with few scales 2–3 mm long; lamina dark green, lanceolate, 1-pinnate, proliferous near apex; pinnae in 10–16 pairs, herbaceous, elongate-deltoid, 6–7.5 × 2–3 cm, basiscopic base obliquely cuneate, acroscopic base parallel to rachis, margin irregularly bidentate, apex attenuate, with minute scattered scales when young but soon glabrescent; basal pinnae hardly smaller, apical pinna ± similar to lateral ones, with gemma at base; veins forked. Sori many, on each side of the costa along the veins at ± 45° angle; indusium membranous, entire. Fig. 2: 6–8, p. 14.

UGANDA. Mengo District: 1.5 km NE of Nansagazi, Sep. 1969, *Faden et al.* 69/1033!
KENYA. Mt Elgon E, July 1944, *van Someren* 622!
TANZANIA. Bukoba District: Minziro Forest Reserve, Feb. 2002, *Bayona & Festo* 92/08!
DISTR. **U** 4; **K** 3; **T** 1; Nigeria, Congo-Kinshasa, Rwanda, Angola
HAB. Moist forest along stream, groundwater forest; 1100–1150 m
CONSERVATION NOTES. Widespread; least concern (LC)

NOTE. *A. gemmascens* does look a lot like *A. warneckei,* apart from being proliferous.
    A specimen that looks very similar is Uganda, **U** 4 Mengo District: Kamengo [Kammengo], July–Aug. 1953, *H.D. van Someren* 772!; except that this is much larger, seems to be shortly creeping rather than long-creeping – and is proliferous in the middle of the costa on the upper surface of many leaflets PLUS near the apex (a character that Faden says is unique to *A. angolense*). Otherwise this specimen looks very like *A. warneckei* (which is non-proliferous, but short-creeping – and known from this area!).

10. **Asplenium sandersonii** *Hook.*, Sp. Fil. 3: 147, t. 179 (1860); Sim, Ferns S. Afr. ed. 2: 139, t. 43/1 (1915); Tardieu in Mém. Inst. Fr. Afr. Noire, 28: 175, t. 33/4–5 (1953); Schelpe, F.Z. Pteridophyta: 183, t. 53b (1970); Burrows, S. Afr. Ferns: 229, map, figs. (1990); Johns, Pterid. trop. East Africa checklist: 67 (1991); Faden in U.K.W.F. ed. 2: 27, t. 172 (1994). Type: South Africa, Natal, Field's Hill, *Sanderson* s.n. (NH , holo.)

Gregarious epiphyte, the frond tips rooting at the tip to form dense colonies; rhizome erect to suberect, ± 3 mm diameter, with tufted fronds and with dark brown lanceolate-attenuate entire clathrate rhizome scales 2–3.5 mm long. Fronds arching, fleshy to coriaceous, 1-pinnate, proliferating with a bud at the end of a naked extension of the rachis; stipe straw-colored, 1–9 cm, sulcate when dry, glabrous except for occasional subulate clathrate scales up to 1.5 mm long. Lamina simply pinnate, 5–30 × 1.5–4 cm, pinnae in 12–23 pairs, shortly petiolate, rhombic-dimidiate to obliquely cuneate tending to half-moon-shaped; pinnae 0.6–2 × 0.3–1 cm, shallowly lobed into 2–9 entire obtuse lobes on the acroscopic margins, subglabrous with occasional minute brown stellate scales ± 0.5 mm diameter on the dorsal surface. Rachis straw-colored, with narrow green wings, set with sparse subulate brown scales and minute paler substellate scales. Sori brown, (1–)2–8 per pinna, oblong, to 3 mm long; indusium oblong, fimbriate, semi-transparent. Fig. 2: 9–10, p. 14.

UGANDA. Kigezi District: Bwindi Forest, Ishasha Gorge, Apr. 1998, *Hafashimana* 506!; Elgon, May 1997, *Wesche* 1339!; Mengo District: Sezibwa [Ssezzibwa] falls 10 km W of Lugazi, Sep. 1969, *Faden & Evans* 69/980!

KENYA. Northern Frontier District: Ndoto Mts, Manmanet ridge, Oct. 1995, *Bytebier & Kirrika* 37!; Kiambu District: Kieni Forest, 8 km E of Kieni, June 1986, *Beentje & Mungai* 2919!; Teita District: Mt Kasigau, path from Rukanga, Apr. 1969, *Faden et al.* 69/470!

TANZANIA. Lushoto District: Kenguruwe–Amani–Sigi Forest, July 1980, *Ruffo & Mmari* 2199!; Morogoro District: Mt Nguru ya Ndege, Sep. 1971, *Pócs & Mwanjabe* 6448L!; Iringa District: Mufindi, Lulanda, Jan. 1989, *Gereau et al.* 2928!

DISTR. U 1–4; **K** 1, 3–7; **T** 1–4, 6, 7; widespread in tropical and South Africa; Madagascar, Comoros

HAB. Moist forest and forest margins, epiphytic on living and dead stems of many species, from ground level up to 7 m high, may be locally common and mat-forming; occasionally on wet rock; (680–)950–2650(–3100) m

CONSERVATION NOTES. Widespread; least concern (LC)

SYN. *Asplenium vagans* Baker, Syn. Fil.: 195 (1867); Alston, Ferns W.T.A.: 59 (1959). Syntypes: Sao Tomé, *Mann* s.n. (? not at K, syn. – see *A. punctatum*, below) and Madagascar, *Meller* s.n. (K!, syn.)
  A. debile Kuhn, Fil. Afr.: 101 (1868), *non* Fée (1865). Type: Comoro Is., *Boivin* s.n. (W, holo.)
  A. melleri Kuhn, Fil. Afr.: 106 (1868). Type: Madagascar, *Meller* s.n. (K!, holo.)
  A. punctatum Kuhn, Fil. Afr.: 114 (1868). Type: Sao Tomé, *Mann* s.n. (K, syn.) – this sheet which has *vagans* Bak. on the cover might be the same as the missing syntype of *vagans*
  A. hanningtonii Baker in Journ. of Bot. 21: 245 (1883). Type: Tanzania, Usafwa Mts, *Hannington* s.n. comm. 1/83 *Mitten* (K!, holo.)
  A. comorense C.Chr., Ind. Fil.: 105 (1906), *nom. nov.* Type as for *A. debile* Kuhn

NOTE. Kuhn describes his *A. debile* as having a creeping rhizome, and does not mention the apex of the lamina; the same goes for his *A. melleri*; both were included in the synonymy of *A. sandersonii* by Schelpe. *A. vagans* Baker was described without having its leaf apex mentioned; it, too, was included under *A. sandersonii* by Schelpe. I have included these provisionally.
  Luke & Luke 5007 from **T** 7, Udzungwa Mts, Sonjo-Mwanihana route, 680 m, is a young sterile plant, but almost certainly this species; it is already budding at several leaf tips. It grows at a much lower altitude than normal in East Africa.

11. **Asplenium monanthes** L., Mant. Pl. 1: 130 (1767); Richter, Codex Bot. Linn.: 1030 (1835); Sim, Ferns S. Afr. ed. 2: 141, t. 46 fig. 1 (1915); Alston, Ferns W.T.A.: 57 (1959); Schelpe, F.Z. Pteridophyta: 175, t. 53d (1970); Burrows, S. Afr. Ferns: 226, map, figs. (1990); Johns, Pterid. trop. East Africa checklist: 66 (1991); Thulin, Fl. Somal. 1: 13 (1993); Faden in U.K.W.F. ed. 2: 28, t. 172 (1994). Type: South Africa, Cape of Good Hope, herb. Linnaeus (LINN 1250/17, lecto.)

Terrestrial, also on dead logs or mossy boulders, rarely a low-level epiphyte; rhizome erect to suberect with tufted fronds and with lanceolate-attenuate fine hair-pointed rhizome scales up to 3.5 mm long, with a black central stripe and paler clathrate entire borders. Frond erect, 1-pinnate, rarely proliferous at top of stipe (not seen in our material). Stipe glossy dark brown to black, 2–20 cm long, glabrous, terete. Lamina pale to dark green, simply pinnate, firmly membranous, linear in outline, 15–60 × 1.8–3.5 cm, decrescent and acute; pinnae up to 45 pairs, oblong-dimidiate becoming broadly cuneate-flabellate towards the base or often the lowermost flabellate, 1–1.7 × 0.5–0.8 cm at about the centre of the lamina, shortly petiolate to subsessile, prominently crenate-dentate on the acroscopic and outer margins, glabrous on both surfaces. Rachis glossy dark brown to black, sulcate on the ventral surface, slightly winged on upper side (always?), the wing green. Sori usually solitary, occasionally 2, rarely with a third on the acroscopic half of the pinna, dark brown, narrowly oblong, to 5 mm long, set near and parallel to the basiscopic margin; indusium whitish, membranous, entire, linear, to 1 mm wide. Fig. 2: 11–12, p. 14.

UGANDA. Karamoja District: Mt Kadam, 1959, *J.Wilson* 784!; Mt Elgon, Benet, Jan. 1936, *Eggeling* 2468! & Suam R., Sep. 1961, *E.J. Brown* EA 12493!
KENYA. Northern Frontier District: Mt Nyiru, Mar. 1995, *Bytebier et al.* 73!; Mt Kenya, Rotundu, Kazita R., *Luke & Luke* 4789!; Masai District: Nsampolai Valley, Sep. 1971, *Greenway & Kanuri* 14903!
TANZANIA. Arusha District: Mt Meru, Jakukumia, Mar. 1971, *Richards* 26853b!; Lushoto District: W Shagai Forest, May 1953, *Drummond & Hemsley* 2726!; Iringa District: Mufindi, near Kigogo R., May 1968, *Renvoize & Abdallah* 1917!
DISTR. **U** 1, 3; **K** 1–7; **T** 2–4, 7; Sudan to South Africa; Madagascar, Réunion, Hawaii and tropical America
HAB. Terrestrial, also on dead logs or mossy boulders in moist forest, especially in bamboo or *Podocarpus* forest, less often in riverine woodland or in the heath zone near streams; may be locally common; 1950–3100(–3400) m
CONSERVATION NOTES. Widespread; least concern (LC)

SYN. *A. monanthemum* L., Syst. Nat. ed. 12, 2: 690 (1767). Type as above

12. **Asplenium normale** *D.Don*, Prodr. Fl. Nepal.: 7 (1825); Schelpe, F.Z. Pteridophyta: 175 (1970); Pic.Serm. in B.J.B.B. 55: 146 (1985); Johns, Pterid. trop. East Africa checklist: 66 (1991); Faden in U.K.W.F. ed. 2: 28 (1994). Type: Nepal, Narainhetty, *Buchanan-Hamilton* s.n. (BM!, holo.)

Terrestrial, very occasionally a low-level epiphyte or lithophyte; rhizome ± 5 mm diameter, erect (?to creeping), with tufted fronds and lanceolate-attenuate subentire clathrate dark brown rhizome scales up to 2.5 mm long, with a narrow ferrugineous margin. Frond erect or arching, narrowly oblong to linear, 15–60 cm long, sometimes proliferous near the apex, firmly membranous. Stipe glossy dark brown, 2–16 cm long, terete, glabrous. Lamina linear in outline, 12–50 × 1.5–3.3 cm, 1-pinnate, hardly decrescent below. Pinnae 18–49 pairs, up to 1.7 × 0.8 cm, shortly petiolate, mostly oblong, slightly dimidiate, rounded, shallowly auriculate, acroscopically becoming increasingly deflexed and unequally triangular towards the base, shallowly crenate on the acroscopic and outer margins, glabrous on both surfaces. Rachis glossy dark brown, sulcate ventrally. Sori brown, up to 3 mm long, 2–7(–8) per pinna, oblong; indusium linear, membranous, entire. Fig. 2: 13–14, p. 14.

UGANDA. Kigezi District: E Virunga, saddle between Muhavura and Mgahinga, Nov. 1954, *Stauffer* 682!; Mt Elgon, Benet, Jan. 1936, *Eggeling* 2469!
KENYA. Kiambu District: Karamenu R. valley, junction of Kitikuyu and Karamenu Rs., Apr. 1971, *Faden et al.* 71/284!; Masai District: Nasampolai Valley, Nov. 1969, *Greenway & Kanuri* 13868!; Teita District: Kasigau, Rukanga route, Nov. 1994, *Luke & Luke* 4121b!
TANZANIA. Lushoto District: Shume-Magamba Forest reserve, May 1987, *Kisena* 530!; Morogoro District: Uluguru Mts, Mwere Valley, Sep. 1970, *Faden et al.* 70/624!; Iringa District: Udzungwa Mt National Park, point 236, Oct. 2001, *Luke et al.* 8070!
DISTR. **U** 2, 3; **K** 3, 4, 6, 7; **T** 2, 3, 6, 7; Malawi, Mozambique; tropical Asia
HAB. Moist forest, where terrestrial or on dead wood, only rarely a low-level epiphyte; occasionally up into the heath zone; 1150–3000 m
CONSERVATION NOTES. Widespread; least concern (LC)

NOTE. Plants of this species resemble the much commoner *A. monanthes* but are usually proliferous; the two taxa are pretty close, and sometimes you have to look at many pinnae before being able to decide (or, just as often, wonder which one). Sterile specimens can be almost impossible to distinguish.

13. **Asplenium gemmiferum** *Schrad.* in Gött. Gel. Anz. 1818: 916 (1818); Alston, Ferns W.T.A.: 56 (1959); Tardieu in Fl. Cameroun Pteridoph.: 188, t. 30/1–2 (1964); Schelpe, F.Z. Pteridophyta: 173 (1970); Burrows, S. Afr. Ferns: 216, map, figs. (1990); Johns, Pterid. trop. East Africa checklist: 63 (1991); Faden in U.K.W.F. ed. 2: 27, t. 172 (1994). Type: South Africa, Cape, ?Grahamstown, *Hesse* s.n. (?LE, holo.)

Terrestrial, epilithic or epiphytic, 60–120 cm tall; rhizome erect, to 30 mm diameter with grey-brown to black, lanceolate acuminate irregularly fimbriate rhizome-scales, 9–14 × 2–3 mm, composed of thin-walled cells. Fronds tufted, arching, somewhat fleshy, to 80 cm long. Stipe matt-grey to brown above and green below, canaliculate, 20–30 cm long and 3 mm diameter, at first densely clothed with subulate clathrate subentire or fimbriate scales, gradually becoming subglabrous. Lamina oblong-lanceolate, 1-pinnate, 40–100 × (14–)20–26 cm, bearing a gemma or small plant at the base of the apical pinna - some fronds lack gemmae but at least some proliferous fronds usually present. Pinnae in 6–12 pairs, alternate, coriaceous when fresh, dark green and shiny above, lanceolate to narrowly elliptic, 10–19 × 2–5.2 cm, petiolate, unequally cuneate at the base with the acroscopic base parallel to the rachis and the basiscopic one cuneate, margin subentire to minutely and regularly shallowly crenate, acuminate, glabrous or nearly so; basal pinnae slightly reduced, terminal pinna similar to other pinnae; veins forked. Rachis matt-grey-green, narrowly winged in the upper $^1/_2$ or not winged, subglabrous as the stipe. Sori many, equally spread along the veins at 45° to midrib, 5–19 mm long, usually extending from near the costa halfway to the margin; indusia linear membranous, entire, ± 1 mm wide. Fig. 3: 1, p. 19.

UGANDA. Toro District: Kibale Forest, Dec. 1957, *Allen* 3720! & Ruwenzori, Mihunga, Jan. 1939, *Loveridge* 343!; Mengo District: Mpanga Forest Reserve 5 km E of Mpigi, Sep. 1969, *Faden et al.* 69/1002!
KENYA. Northern Frontier District: Mt Marsabit, June 1960, *Oteke* 30!; Meru District: Nyambeni Hills, R. Thangatha, Oct. 1960, *Polhill & Verdcourt* 271!; Masai District: Chyulu Hills, main forest N, Dec. 1993, *Luke & Luke* 3879!
TANZANIA. Kilimanjaro, above Kilimanjaro Timbers, June 1993, *Grimshaw* 93/144!; & Kidia, Old Moshi, Msaranga valley, Mar. 1997, *Hemp* 1633!; Iringa District: Udzungwa Mountain National Park, Sonjo–Mwanihana route, Nov. 1997, *Luke & Luke* 5053a!
DISTR. **U** 2–4; **K** 1, 3, 4–7; **T** 2, 3, 6, 7; Bioko, Cameroon, Congo-Brazzaville, Congo-Kinshasa, Burundi, Malawi, Mozambique, Zimbabwe, South Africa
HAB. Moist forest, terrestrial, on rocks or a low epiphyte; 1050–2250 m
CONSERVATION NOTES. Widespread; least concern (LC)

SYN. *A. macrolobium* Peter in F.D.O.-A.: 77 (1929) & Descr. 6, t. 3/1–2 (1929); Johns, Pterid. trop. East Africa checklist: 65 (1991). Type: Tanzania, N Pare Mts, Kilomeni to Kissangara, *Peter* 11579 (B!, holo.; K!, iso.)

NOTE. This species can be confused with *boltonii* (but has entire or crenate pinnae), and *anisophyllum* (but has gemmae near apex).
　　Luke & Luke 5053a has only few sori, near the base of each pinna. *Grimshaw* notes that above 2100 m on Kilimanjaro this taxon is replaced by *Asplenium* sp. B of Johns, = *A. boltonii*, with which it seems to intergrade.

14. **Asplenium adamsii** *Alston* in Bol. Soc. Brot. 30: 7 (1956) & Ferns W.T.A.: 59 (1959); Tardieu, Fl. Cameroun Pterid.: 210, t. 25/3–5 (1964); Johns, Pterid. trop. East Africa checklist: 61 (1991); Faden in U.K.W.F. ed. 2: 28 (1994). Type: Cameroon, Mt Cameroon, *Adams* 1278 (BM!, holo.)

Terrestrial or lithophyte; rhizome shortly creeping (as in the type!) or erect, ± 5 mm diameter, with dense shiny clathrate black scales 2–2.5 × 0.8 mm long, margin paler, apex ending in a hair. Fronds tufted, oblong-elliptic, 20–36 cm long, sometimes proliferous. Stipe green to blackish, 6–14 cm long, with some scales similar to those of rhizome and a few capitate glands. Lamina oblong in outline, 10–22 × 2.5–6 cm,

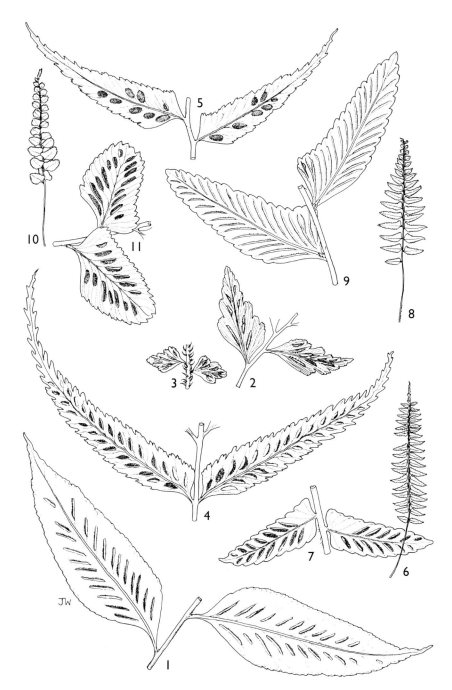

FIG. 3. *ASPLENIUM GEMMIFERUM* — **1**, pinnae × ²/₃. *ASPLENIUM ADAMSII* — **2**, pinnae × 1; **3**, young pinnae × 1. *ASPLENIUM BOLTONII* — **4**, pinnae × ²/₃. *ASPLENIUM ELLIOTII* — **5**, pinnae × ²/₃. *ASPLENIUM BARTERI* — **6**, habit, not to scale; **7**, pinnae × 1. *ASPLENIUM MACROPHLEBIUM* — **8**, habit, not to scale; **9**, pinnae × 1. *ASPLENIUM CHRISTII* — **10**, habit, not to scale; **11**, pinnae × 1. 1, from *J.Beesley* 132; 2, 3, from *Faden* 69/155; 4, from *Faden et al* 272a; 5, from *Faden & Cameron* s.n.; 6, 7, from *Hafashimana* 420; 8, from *Faden* 69/2118; 9 from *Jonsson* 3871/2; 11, 12, from *Sir John Kirk* 8/1884. Drawn by Juliet Williamson.

1-pinnate, basal pinnae hardly reduced, apex gradually decrescent; pinnae in 9–19 pairs, sub-coriaceous, to 4 × 1.3 cm, base unequally cuneate-attenuate, the acroscopic margin parallel with the rachis, the basiscopic margin slightly dimidiate and then cuneate, sometimes slightly 3-lobed, upper margin double serrate, apex obtuse to attenuate. Rachis with scattered scales similar to those of rhizome, and with narrow hairs/scales, some of which are small stalked glands. Sori 6–17 per pinna, at slight angle to costa, linear, 2.5–6 mm long; indusium membranous, entire, to 0.7 mm wide. Fig. 3: 2–3, p. 19.

KENYA. Elgon, 11200 feet, Apr. 1935, *G. Taylor* 3735!; North Nyeri District: Nyandarua/Aberdare Mts, Cave waterfall, July 1960, *Polhill* EA 12026!; idem, Feb. 1969, *Faden* 69/155!
TANZANIA. Arusha District: Mt Meru crater, Apr. 1968, *Vesey-Fitzgerald* 5632! & idem, north wall, Apr. 1969, *Vesey-Fitzgerald* 6175!
DISTR. **K** 3, 4; **T** 2; Cameroon
HAB. Rocky places near waterfall in moorland and upper heath zone, crevices in lava wall; 2400–3400 m
CONSERVATION NOTES. Due to wide distribution and habitat type, presumed Least Concern (LC)

15. **Asplenium boltonii** *Schelpe* in Bol. Soc. Brot. ser. 2, 41: 204 (1967) & F.Z. Pteridophyta: 173 (1970); Pic.Serm. in Webbia 37, 1: 133 (1983); Burrows, S. Afr. Ferns: 216, map, figs. (1990); Johns, Pterid. trop. East Africa checklist: 62 (1991); Faden in U.K.W.F. ed. 2: 28 (1994). Type: South Africa, Natal, *Bolton* s.n. (K!, holo., BM, iso.)

Low-level epiphyte, less often lithophyte or terrestrial; rhizome erect or sometimes shortly creeping, fleshy, up to 2 cm diameter, branching on large plants, with brown linear-lanceolate scales with pale margins, entire, up to 18 × 2 mm at its base, composed of hyaline thin-walled cells. Fronds tufted, arching, 40–100 cm high, 1-pinnate, proliferous at the base of one of the apical pinnae. Stipe matt-grey-green, 10–26 cm long and 4 mm diameter, at first densely covered with reddish-brown hair-like scales up to 14 mm long, later becoming subglabrous. Lamina dull green, chartaceous, 1-pinnate, ovate-lanceolate, 33–80 × 8–22 cm, with the lowest pinnae slightly reduced. Pinnae in 13–28 pairs, lanceolate-attenuate, 4–10(–15) × 1–1.5(–1.8) cm, base broad and unequally cuneate, the acroscopic part parallel to the rachis, margin incised ¹⁄₃ way to the costa into bilobed or bicrenate-serrate lobes, apex attenuate, glabrous above but with scattered hair-like scales below when young; veins forked. Rachis matt-grey-green, with some hair-like scales similar to those on the stipe. Sori several to many, elliptic when mature, 3–8 mm long, extending along the veins from near the costa to ¹⁄₂ or ²⁄₃ the distance to the margin; indusium subelliptic, 1–1.5 mm broad, entire. Fig. 3: 4, p. 19.

UGANDA. Ankole District: Kalinzu forest, 4 km NW of sawmill, Sep. 1969, *Faden et al.* 69/1160!; Kigezi District: Bwindi forest, Ihihizo, Aug. 1998, *Hafashimana* 779!; Mengo District: Kiwala, Apr. 1916, *Dummer* 2803!
KENYA. Fort Hall District: Kimakia Forest Station, 8 km E on Njabini road, July 1969, *Faden & Evans* 69/895!; Kericho District: 5 km E of Kericho along Timbilil R., June 1972, *Faden et al.* 72/306!; Teita District: Taita Hills, Ngangao forest, May 1985, *NMK Taita Hills Expedition* 312!
TANZANIA. Kilimanjaro, above Mandera Hut, Oct. 1993, *Grimshaw* 93/868!; Morogoro District: Mwanihana Forest reserve, Oct. 1984, *D.W. Thomas* 3878!; Njombe District: Livingstone Mts, Madunda Mission, Feb. 1961, *Richards* 14095!
DISTR. **U** 2, 4; **K** 3–5, 7; **T** 2, 6, 7; Congo-Kinshasa, Malawi, Zambia, Mozambique, Zimbabwe, South Africa; Madagascar, Réunion
HAB. Moist forest; 1200–2750 m
CONSERVATION NOTES. Widespread; least concern (LC)

SYN. *A. anisophyllum* Kunze var. *b* ; Kunze in Linnaea 10: 512 (1836). Type: South Africa, near Philipstown at source of Kat River, June, *Ecklon* s.n. (herb. Kunze; not found at K, possible specimen at B sub *Ecklon* 30)

A. anisophyllum Kunze var. *elongatum* Mett. in Abh. Senckenb. Naturf. Ges. 3: 143 (1859) reimpr. in Mett., Farngatt., 6: 99 (1859) based on *Asplenium anisophyllum* Kunze var. b
A. anisophyllum Kunze var. *latifolium* Hook., Sp. Fil. 3: 111, t. 166 (1860) based on *Asplenium anisophyllum* Kunze var. b
Asplenium anisophyllum sensu Sim, Ferns S. Afr. ed. 2: 151, t. 53 (1915); Tardieu, Fl. Madag.: 203, t. 27 fig. 9–10 (1958), *non* Kunze
Asplenium sp. B of Johns, Pterid. trop. East Africa checklist: 68 (1991)

NOTE. There has been confusion over the authority of the name, but despite Pichi Sermolli's arguments (loc.cit.) I believe Schelpe was the first to publish the name correctly. Brause & Hieron. in Z.A.E.: 8 (1910) merely repeated the non-publication of the name in Baker's Syn. Fil.: 204 (1874).

I agree with Faden that this resembles the much commoner *A. smedsii* and that it differs only in having gemmae on the fronds. Burrows (loc. cit.) states that the occurrence of gemmae on the fronds varies in different areas. This is certainly true for East Africa, at the norther end of the species' range. In quite similar-looking plants from Ngangao forest, **K** 7, gemmae may be present or absent; Faden states in his 69/895 from **K** 4 that the whole population lacked gemmae – possibly the reason why Johns in his checklist split off this collection and *Verdcourt* 2988 as 'species B'. Relationships between these two taxa (*boltonii* and *smedsii*) need to be cleared up.

In the Berlin herbarium there were several sheets of this taxon named *A. anisophyllum* Kunze var. *microphylla*, a name on which I have been unable to find any information. Similarly specimens of *A. boltonii* were named *A. deckenii* Hieron. with as 'type' a specimen from Tanzania, Kilimajaro, Dschagga, 5500'–7800', leg. Kersten anno 1864 in *von der Decken* 27 (and possibly 28). Again I have found no publications in which this name is mentioned.

16. **Asplenium elliottii** *C.H.Wright* in K.B. 1908: 262 (1908); Johns, Pterid. trop. East Africa checklist: 63 (1991); Schippers in Fern Gaz. 14, 6: 200 (1993); Faden in U.K.W.F. ed. 2: 28, t. 172 (1994). Type: Kenya, Aberdare range at ± 2800 m, *C.F. Elliott* s.n. (K!, holo.)

Terrestrial or less often a low-level epiphyte, often gregarious, 30–140 cm high; rhizome erect (very rarely shortly creeping), to 8 mm diameter, with concolorous ovate scales 4–10 × 2.5–4 mm. Stipe straw-colored, 11–30(–80) cm long, thin, glabrous or with sparse subulate clathrate scales to 4 mm long with widened base and occasionally with very thin side-lobes. Fronds tufted. Lamina dark green, lanceolate, 24–60(–100+) × 10–22 cm, 1-pinnate, the lowermost pinnae slightly smaller, decrescent towards the apex, proliferous near apex; pinnae in 14–30 pairs, opposite or alternate, rhomboid or narrowly lanceolate, (3–)5–16(–22) × (1.3–)1.6–3.3 cm, base truncate and parallel to rachis on the acroscopic side, oblique on the basiscopic side, minutely crenate-serrate, acuminate; veins forked once; glabrous or occasionally with sparse scattered scales; petiolule 1–4 mm long. Rachis slightly winged in the distal part, with sparse subulate scales to 3.5 mm long with widened base and occasionally with very thin side-lobes. Sori many, from near costa to halfway towards the margin, brown or red-brown, linear to elliptic, 3–4.5(–8) mm long; indusium pale, entire, 0.5–1(–1.3) mm wide. Fig. 3: 5, p. 19.

UGANDA. Toro District: Ruwenzori, Bjuku valley, Jan. 1951, *Osmaston* 3662!; Kigezi District: Bwindi forest, Rubanda, Nov. 1989, *Rwaburindore* 2886!; Elgon, Dec. 1996, *Wesche* 573!
KENYA. Meru District: Nyambeni Hills, Kirima, Oct. 1960, *Verdcourt & Polhill* 2973!; Kericho District: Chemasingi Tea Estate 13 km S of Kericho, Dec. 1967, *Perdue & Kibuwa* 9304!; Teita District: Taita Hills, Ngangao, May 1985, *NMK Taita Hills Expedition* 272!
TANZANIA. Lushoto District: Baga Forest Reserve, May 1987, *Kisena* 635!; Ufipa District: Nsanga forest, Aug. 1960, *Richards* 12970!; Mpwapwa District: Sagara [Usagura] Mts, 1884, *Kirk* s.n.!
DISTR. **U** 2, 3; **K** 3–7; **T** 1–7; Congo-Kinshasa, Rwanda, Burundi
HAB. In moist montane forest, extending up into the bamboo zone; may be locally common to abundant; 1050–2800 m
CONSERVATION NOTES. Widespread; least concern (LC)

SYN. *A. anisophyllum* Kunze var. *aequilateralis* Hieron. in P.O.A. C: 82 (1895). Type: Tanzania, Lushoto District: Usambara, Shagayu Forest near Mbaramu, *Holst* 2491 (B!, holo.)

A. *anisophyllum* Kunze var. *pseudo-plumosa* Hieron., P.O.A. C: 82 (1895). Type: Tanzania, Lushoto District: Usambara Mts, Shagayu forest near Mbaramu, *Holst* 2492 (B!, holo.), **syn. nov.**

A. *aequilaterale* (Hieron.) Viane in Biol. Jaarb. 59: 157 (1991); Schippers in Fern Gaz. 14, 6: 198 (1993); Johns, Pterid. trop. East Africa checklist: 61 (1991), **syn. nov.**

NOTE. Several specimens from **K** 7, Shimba Hills, Mar. 1941, *H.D. van Someren* 125!, 151! and 168! are probably this taxon, but lack rhizomes. This would be at much lower altitude than other specimens, though the labels lack altitude data.

A specimen from Uganda, Ruwenzori, Mobuku valley, Dec. 1938, *Loveridge* 291! looks very much like this taxon; but is strikingly different from all other East and central African *Asplenium* species I have seen in that many a pinna has on the upper surface, in the middle of the costa, a gemma with growing plantlet. Most details are lacking: the specimen I have seen consists of a 33 cm part of rachis with 9 pairs of what looks like upper-middle pinnae. The fern is said to be 2 m high – unknown whether creeping or tufted. Presumably 1-pinnate, the pinnae 9 or more (presumably many more) on each side of the rachis; pinnae lanceolate, to 15 × 2.6 cm, shortly stalked, acroscopic base parallel to the rachis and almost with an enlarged lobe, basiscopic base cuneate, margin crenate, apex acuminate, glabrous or nearly so. Rachis with very sparse filiform scales to 2 mm long. Sori many, restricted to the lower half of the pinna, at a 45° angle to the costa and slightly nearer the costa than the margin, linear, 3–5 mm long, straight; indusium translucent, to 1 mm wide, entire.

The name A. *pseudoplumosum* Hiern, under which several specimens were laid in at Berlin, was never published.

17. **Asplenium barteri** *Hook.*, Sec. Cent. Ferns 2: t. 75 (1861); Tardieu, Mém. I.F.A.N. 28: 180, t. 34/2 (1953); Alston, Ferns W.T.A.: 56 (1959); Tardieu, Fl. Cameroun Pterid.: 192 (1964); Johns, Pterid. trop. East Africa checklist: 62 (1991). Type: Nigeria, Aboh, *Barter* 1454 (K!, holo., B!, iso.)

Epiphyte; rhizome erect, short, with dark brown narrowly triangular acuminate scales to 4 mm, with pale brown margin with few hair-like lobes. Fronds tufted, proliferous. Stipe greyish, 10–30 cm long, with few subulate attenuate scales to 3 mm. Lamina ovate-lanceolate, 14–30 × 4–6 cm, 1-pinnate, basal pinnae not or only slightly reduced, apical pinnae decrescent to a linear lobed terminal pinna, at its base often proliferous. Pinnae in 13–18 pairs, opposite or alternate, close, subcoriaceous, ovate or elliptic, 2–4 × 0.5–1.2 cm, sessile, base unequal, the acroscopic base parallel to the rachis and often slightly auriculate, the basiscopic base obliquely cuneate, margin slightly crenate-dentate, apex short-acuminate; veins simple (or the basal one forked). Rachis scaly. Sori in 6–7 pairs on each side of the costa, at angle of 45°, 2–4 mm long; indusium linear, membranous, entire, 0.4–0.5 mm wide. Fig. 3: 6–7, p. 19.

UGANDA. Kigezi District: Ishasha Gorge, Nov. 1946, *Purseglove* 2255! & idem, 7 km SW of Kirima, Sep. 1969, *Faden et al.* 69/1214! & Ishasha Gorge, Nov. 1997, *Hafashimana* 426!
DISTR. **U** 2; **K** 7 (see Note); **T** 1 (see Note); from Guinea to Congo-Kinshasa
HAB. Low epiphyte on tree trunks; moist forest, close to river; 1200–1350 m
CONSERVATION NOTES. Widespread; least concern (LC)

NOTE. Three specimens from Kenya **K** 7, Taita Hills: Mbololo, collected on the *NMK Taita Hills Expedition* (nrs. 778A!, 782! and 1126!) key out here but differ in being terrestrial or lithophytic. Rhizome scales in these three specimens are almost absent but the remnants look quite similar. As the distribution area is remote, they are mentioned here tentatively. A *Holtz* specimen, Bukoba District: Minziro Forest Reserve, July 1906, *Holtz* 1708! looks quite like A. *barteri* but has a rather short stipe (7 cm long), and sori to 6 mm long; there are very few rhizome scales; it is somewhat intermediate between *barteri* and *macrophlebium*.

18. **Asplenium macrophlebium** *Baker* in Hooker & Baker, Syn. Fil.: 485 (1883); Alston, F.W.T.A.: 56 (1959); Tardieu, Fl. Cameroun Pteridophytes: 193 (1964); Johns, Pterid. trop. East Africa checklist: 65 (1991); Schippers in Fern Gaz. 14, 6: 201 (1993); Faden in U.K.W.F. ed. 2: 28, t. 172 (1994). Type: Bioko [Fernando Po], *Mann* 338 (K!, holo.)

Terrestrial or rarely low-level epiphyte; rhizome erect, less often creeping, short, with pale brown lanceolate subentire clathrate scales to 3 mm long. Fronds tufted (once descibed as spaced on a long-creeping rhizome, fide *Faden*), oblong-lanceolate, 15–60 × 3–9 cm, usually proliferous at apex. Stipe green, canaliculate, 5–16 cm long, with narrow lanceolate scales. Lamina subcoriaceous, 1-pinnate, the basal ones not decreasing in size, the middle ones opposite or alternate, decrescent towards apex. Pinnae in 7–20 pairs, subsessile, at right angles to rachis, lanceolate, 2–6 × 0.9–2.3 cm, base unequal, the basiscopic part cuneate, the acroscopic part cuneate and parallel with the rachis for a bit and there forming a lobe, margin deeply crenate, apex obtuse; costa whitish, veins forked in and near acroscopic basal lobe but otherwise unbranched. Rachis flattened, narrowly winged upwards, slightly scaly. Sori elongate, many, closely parallel, almost reaching costa and margin, 2–10 mm long; indusium membranous, entire or nearly so, narrow, to 0.5 mm wide. Fig. 3: 8–9, p. 19.

UGANDA. Bunyoro District: Budongo forest, Nyakafunjo Nature Reserve, Nov. 1996, *Hafashimana* 99!; Kigezi District: Ishasha Gorge, Nov. 1946, *Purseglove* 2256!; Mengo District: Sezibwa [Ssezibwa] Falls 10 km W of Lugazi, Sep. 1969, *Faden & Evans* 69/972!
KENYA. N Kavirondo District: Kakamega Forest NE of Forest Station, Nov. 1969, *Faden et al.* 69/1999!; Nandi District: between Kaimosi Tea Estates and Yala R., Dec. 1969, *Faden & Rathbun* 69/2118!; Teita District: Taita Hills, Ngangao Forest, Feb. 1977, *Faden & Faden* 77/323!
TANZANIA. Bukoba District: Minziro Forest reserve, July 2000, *Bidgood et al.* 4777!; Morogoro District: Uluguru Mts, Mwere valley, Sep. 1970, *Faden et al.* 70/617!; Iringa District: Udzungwa Mountain National Park, Mt Luhomero, Sep. 2000, *Luke et al.* 6739!
DISTR. U 2–4; K 3, 5, 7; T 1, 4, 6, 7; West Africa from Guinea to Cameroon, Bioko, Congo
HAB. Moist forest, where may be locally common; 1000–1850 m
CONSERVATION NOTES. Widespread; least concern (LC)

NOTE. The material from the Taita Hills (**K** 7) and the Uluguru Mts (**T** 6) differs in that the sori do not stretch from very close to the costa to nearly reaching the margin – in these specimens they only occupy the upper $^2/_3$ of the vein; also, the indusia are wider, to 0.8 mm, with wavy margin, and the stipe is slightly more scaly than in more Western specimens. The single specimen from the Udzungwa Mts (**T** 7), however, is more like the Western populations! I have decided against subspecific status, but this was a fairly close call.

19. **Asplenium christii** *Hieron.* in P.O.A. C: 82 (1895); Schelpe, F.Z. Pteridophyta: 172 (1970); Johns, Pterid. trop. East Africa checklist: 62 (1991); Burrows, S. Afr. Ferns: 220, map, figs. (1990); Faden in U.K.W.F. ed. 2: 28 (1994). Type: Tanzania, Lushoto District: Usambara Mts, Silai, *Holst* 2304 (B!, holo.; K!, iso.)

Terrestrial or less often low epiphyte or lithophyte; rhizome ± 4 mm diameter, erect or ascending, with concolorous brown narrowly triangular entire scales up to 3 × 0.5 mm at the base. Fronds tufted, dimorphic or uniform, fertile fronds proliferous, 1.5–2 times as long as the sterile non-proliferous fronds. Stipe matt-grey-green, 5–12 (sterile) or 6–26 cm (fertile) long, 1–1.5 mm diameter, at first with sparse lanceolate-acuminate scales ± 1 mm long, similar to those on the rhizome, later becoming subglabrous. Lamina thinly coriaceous, 1-pinnate; sterile lamina broadly lanceolate to ovate, 6–15 × 3–8 cm, pinnae closely spaced and not proliferous. Fertile lamina narrowly lanceolate to lanceolate, 13–27 × 5–8.5 cm, widely spaced and proliferous below the terminal pinna or occasionally on the pinna midrib. Pinnae in 5–12 pairs, alternate, subsessile, asymmetrically ovate to lanceolate and often slightly falcate, 2–6 × 1–2.3 cm, base unequally cuneate, the lower pinnae often with a pronounced lobe on the acroscopic side, margin serrate to sharply crenate, apex acute to attenuate, glabrous except for a few small brown scales along the costa below; veins forked; lower pinnae similar to others; pinnae decreasing in size towards the apex of the frond, terminal pinna similar to other pinnae, bearing a bud at its base in fertile fronds. Rachis similar to stipe. Sori several to many, along the veins from near the costa to halfway to the margin, linear, 3–8 mm long, indusium entire, 0.4–0.8 mm wide, membranous. Fig. 3: 10–11, p. 19.

UGANDA. Toro District: Kibale forest, July 1938, *A.S. Thomas* 2294!; Ankole District: Kalinzu forest near Kyamahunga, Jan. 1953, *Dawkins* 769!; Elgon, Mbale, Dec. 1950, *H.D. van Someren* 566!
KENYA. N Kavirondo District: Kakamega forest, Kibiri block, Mar. 1977, *Faden & Faden* 77/920!; Teita District: Ngangao forest, Feb. 1977, *Faden & Faden* 77/321! & *Taita Hills Expedition* 261!
TANZANIA. Lushoto District: W Usambara, Kwagororo forest, Mar. 2003, *Hemp* 3615! & Amani area, June 1970, *Faden* 70/283!; Morogoro District: Uluguru Mts, Morningside–Bondwa road, July 1970, *Faden, Evans & Kabuye* 70/307!
DISTR. **U** 2–4; **K** 3, 5, 7; **T** 3, 6, 7; Mozambique, Zimbabwe, South Africa
HAB. Moist forest, where it may be common; (300–)900–2000(–2300) m
CONSERVATION NOTES. Widespread; least concern (LC)

SYN. *A. amaurophyllum* Peter, F.D.O.-A.: 73 (1929) & Descr.: 5, t. 1.11–1.12 (1929). Type: Tanzania, Lushoto District: E Usambara, near Amani, *Peter* 10211 (B!, holo.)
   A. amaurophyllum Peter forma *major* Peter, F.D.O.-A. Descr.: 6 (1929). Type: Tanzania, Bomole near Amani, *Peter* 21602 (not found)
   A. amaurophyllum Peter var. *longicuspe* Peter, F.D.O.-A. Descr.: 6, t. 1.13 (1929). Type: Tanzania, Amani to Monga, *Peter* 301g (not found)

NOTE. This species rather resembles small plants of *A. elliottii*. The similar *A. mossambicense* Schelpe is only known from two sites in Zimbabwe and Mt Gorongosa in Mozambique.

20. **Asplenium megalura** *Hieron.* in Z.A.E. 2: 17 (1910); Tardieu in Mém. I.F.A.N. 28: 190, t. 36/2 (1953); Alston, Ferns W.T.A.: 57 (1959); Tardieu, Fl. Cameroun, Ptérid.: 208, t. 31/9 (1964); Schelpe, F.Z. Pteridophyta: 180 (1970); Johns, Pterid. trop. East Africa checklist: 65 (1991); Faden in U.K.W.F. ed. 2: 28, t. 172 (1994). Syntypes from Tanzania: Kwai, *Albers* 288 (B!, syn.); *Buchwald* 136 (B!, K!, syn.); Usambara, *Busse* 380 (B!, syn.), 1145; Kwai, *Eick* 413 (B!, syn.); Amani, *Engler* 573 (B!, syn.); Bulua, Mpinga, *Holst* 83 (B!, syn.), 4255a (B!, syn.), 9145 (B!, syn.); Uluguru, *Stuhlmann* 8902 (B!, syn.); Wuga to Manka, *Uhlig* 1504 (B!, syn.); *Volkens* 61 (B!, syn.) and Congo, Lake Kivu, *Mildbraed* 1235 (B!, syn.); lectotype, chosen here, Tanzania: Kwai, *Buchwald* 136 (B!, K!, lecto.)

Epiphyte, rarely lithophyte or terrestrial; rhizome erect or shortly creeping, to 5 mm diameter, with dark brown narrowly lanceolate attenuate entire rhizome scales up to 6 mm long with narrow pale brown borders, ending in hair-tip. Frond tufted, rarely shortly spaced, arching or drooping, firmly membranous, not proliferous. Stipe shiny dark brown or greyish, 7–19 cm, long, glabrous except at the very base. Lamina oblong, 15–43 × 7–13(–16) cm, 1-pinnate, lowest pinnae not reduced, apical segment resembling others, often tricuspidate. Pinnae in 4–15 pairs, (sub-)opposite, subsessile, wedge-shaped (rhomboid-trapeziform), 3-lobed, up to 9 × 4 cm, up to 11 pairs, base cuneate, lower margin entire, apex very longly caudate-cuneate, the distal margin irregularly serrate-dentate and the cauda also serrate, glabrous on both surfaces or with some scattered tiny lobed scales to 0.5 mm, veins flabellate, costa not clear. Rachis shiny dark brown, shallowly channelled ventrally, glabrous except for a few very small scales. Sori many per pinna, along veins, linear, (3–)5–20 mm long; indusium linear, membranous, entire, 0.4–0.5 mm wide. Fig. 4: 1–2, p. 25.

UGANDA. Toro District: Kibale National Park, Kanyawara, Sep. 1997, *Hafashimana* 331!; Kigezi District: Bwindi Impenetrable National Park, Rukingiri, Aug. 1998, *Hafashimana* 778!; Mbale District: Bulago, Aug. 1932, *A.S. Thomas* 326!
KENYA. Trans Nzoia District: Cherangani, Kapolet [Kabolet], Aug. 1963, *Tweedie* 2683!; Kericho District: 8 km ENE of Kericho, Kitenges R. crossing, June 1972, *Faden et al.* 72/300!; Teita District: Mt Kasigau, above Rukanga, Feb. 1971, *Faden et al.* 71/137!
TANZANIA. Lushoto District: Gare, Mar. 1982, *Mtui & Kisena* 116!; Mpanda District: Mahale Mts, Kasangazi, July 1958, *Juniper et al.* 261a!; Iringa District: Ihangana Forest Reserve near Kibengu, Feb. 1962, *Polhill & Paulo* 1532!
DISTR. **U** 2, 3; **K** 4, 5, 7; **T** 3, 4, 6, 7; tropical Africa from Sierra Leone to Cameroon, Malawi, Mozambique, Zambia
HAB. Moist forests, often near streams; 850–2200 m
CONSERVATION NOTES. Widespread; least concern (LC)

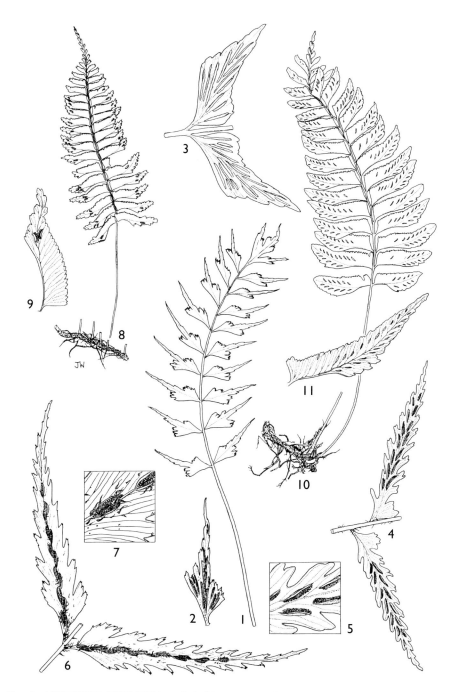

FIG. 4. *ASPLENIUM MEGALURA* — **1**, habit, upper surface × $^2/_3$; **2**, pinna with sori × $1^1/_2$. *ASPLENIUM WARNECKEI* — **3**, pinnae × $^2/_3$. *ASPLENIUM PELLUCIDUM* — **4**, pinnae × $^2/_3$; **5**, sori × 2. *ASPLENIUM FRIESIORUM* — **6**, pinnae × $^2/_3$; **7**, sori × $1^1/_2$. *ASPLENIUM UNILATERALE* — **8**, habit, lower surface × $^1/_3$; **9**, pinna × 1. *ASPLENIUM OBSCURUM* — **10**, habit, undersurface with sori × $^1/_3$; **11**, pinna × $^2/_3$. 1, 2, from *F. Rose* 10300; 3, from *Hafashimana* 0029; 4, 5, from *Luke* 5249; 6, 7, from *Cameron* 144; 8, 9, from *Poulsen* 890; 10, 11, from *Faden et al.* 96/67. Drawn by Juliet Williamson.

SYN. *A. dimidiatum* Sw. var. *longicaudatum* Hieron. in E.J. 28: 343 (1900). Type: Tanzania,
   Uluguru near Kibango, *Stuhlmann* 8902 (B!, holo.)

NOTE. Faden reports that old stipes and rachises may persist.

21. **Asplenium warneckei** *Hieron.* in E.J. 46: 367 (1911); Alston, Ferns W.T.A.: 59
(1959); Tardieu, Fl. Cameroun Pteridophytes: 204, t. 32/6–7 (1964); Johns, Pterid.
trop. East Africa checklist: 68 (1991). Type: Tanzania, Usambara, Amani, *Warnecke*
338 (B!, holo.; B!, K!, P!, iso.)

   Terrestrial or epiphyte; rhizome short-creeping or short-ascending, to 5 mm thick,
with black narrowly triangular rhizome scales to 3.5 × 0.4–0.5 mm, with very thin brown
margin, attenuate into hair-tip. Fronds almost tufted to shortly spaced, 30–80 cm long,
not proliferous (rarely proliferous near apex). Stipe grey or black, green on upper
surface near apex, 10–35 cm long, at base with scales like rhizome scales, otherwise
with brown sparse scales to 2.5 mm. Lamina oblong, 20–45 × 8–18 cm, 1-pinnate, the
lower pinnae rarely sub-3-lobed or lobed at base, the terminal pinna often 3-lobed and
acuminate. Pinnae in 5–9 pairs, patent or spreading, (sub)opposite, chartaceous,
glaucous green, shortly stalked, inaequilateral, trapezo-rhomboid, to 10 × 4.8 cm,
acroscopic base subtruncate-cuneate, basiscopic base cuneate, margin irregularly
serrate-crenate, glabrous or nearly so. Rachis black with thin green lines connecting
pinna bases, with dense scales when young, later with scattered subulate scales. Sori
9–18 per pinna, on the veins at a very slight angle to costa, 0.5–3 cm long, not reaching
margin; indusium membranous, entire, 0.7–0.8 mm wide. Fig. 4: 3, p. 25.

UGANDA. Bunyoro District: Budongo forest, Kaniyo–Pabidi beat, Feb. 1996, *Hafashimana* 29!;
   Mengo District: 14 km from Kampala on Masaka road, May 1937, *Chandler* 1625! & Sezibwa
   [Ssezzibwa] Falls, 10 km W of Lugazi, Sep. 1969, *Faden & Evans* 69/982!
TANZANIA. Lushoto District: E Usambara, Ngambo to Kwamkuyu, Apr. 1915, *Peter* 10000!;
   Morogoro District: North Uluguru Forest Reserve, Mnavu, Dec. 1993, *Kisena* 988!; Iringa
   District: Udzungwa Mountain National Park, Oct. 2001, *Luke et al.* 8158!
DISTR. **U** 2, 4; **T** 1, 3, 6, 7; Ghana, Cameroon
HAB. Moist forest near water; 900–1350 m
CONSERVATION NOTES. Widespread; least concern (LC)

SYN. *A. warneckei* Hieron. var. *prolifera* Hieron. in E.J. 46: 369 (1911). Type: Tanzania, Usambara,
   Lutindi, anno 1900, *Libusch* s.n. (B!, holo.), **syn. nov.**

NOTE. Very close to *A. hemitomum*, from West and central Africa, but distinct in the size of the
   rhizome scales (longer in *hemitomum*), the texture of the leaf (more leathery in *hemitomum*),
   and spores: irregularly reticulate in *hemitomum*, crested in *warneckei*. This taxon (*warneckei*)
   also looks like *A. macrophyllum* from Asia, and Peter cited his number 10000 (cited above) as
   such, but *warneckei* has consistently wider pinnae and a different feel to it. I prefer to keep
   it separate.
   I have made var. *prolifera* into a synonym, though Hieronymus felt it differed in stipe
   colour, dissection of pinnae margins, and being proliferous near the apex; as the type is
   sterile and lacks basal plant, it is not possible to be entirely certain, but I believe this is only
   a form. Two other specimens are proliferous: Uganda, Mengo District: Kipayo, Jan. 1914,
   Dummer 585! & 15 km from Kampala on Mascha road, May 1937, *Chandler* 1625!
   A specimen from **T** 3, Zaraninge forest, collected in 1990 by *Frontier Tanzania* 1099! (MO)
   looks superficially like this taxon, but is from 300 m altitude, has longer rhizome scales and
   narrower pinnae. It might represent a new taxon.

22. **Asplenium pellucidum** *Lam.*, Encycl. Méth., Bot. 2: 305 (1786); Johns, Pterid.
trop. East Africa checklist: 66 (1991). Type: Mauritius [Ile de France], *Commerson* s.n.
(P!, holo.)

   Lithophytic and low-level epiphyte; rhizome erect to occasionally creeping (*Faden*
70/286), to 10 mm thick, with dense subulate brown clathrate attenuate rhizome
scales up to 12 mm long with a few thread-like side lobes. Fronds tufted to rarely

spaced (*Faden* 70/286), arching, not proliferous, thinly coriaceous. Stipe matt-brown, up to 18 cm long, densely covered with subulate hair-like scales similar to those on the rhizome, up to 8 mm long. Lamina lanceolate, 22–100 × 9–25 cm, 1-pinnate to 2-pinnatifid or 2-pinnatipartite on lower pinnae, acute, basal pinnae reduced, apical ones decrescent. Pinnae 20–36 pairs, opposite or alternate, lanceolate, 5–8.5 × 0.8–1.7 cm, base unequally cuneate, slightly auriculate acroscopically, shallowly incised into obtuse or shallowly crenate lobes, apex attenuate, sparsely set with dark hair-like and ovate long-acuminate scales up to 3 mm long or subglabrous; veins forked. Rachis with scales as on stipe. Sori linear, 4–6 mm long, set at about 15° to the costa; indusium linear, membranous, entire, 0.8 mm wide. Fig. 4: 4–5, p. 25.

subsp. **pseudohorridum** (*Hieron.*) *Schelpe* in Bol. Soc Brot., Sér. 2, 41: 208 (1967) & F.Z. Pteridophyta: 179 (1970); Burrows, S. Afr. Ferns: 230, map, figs. (1990); Schippers in Fern Gaz. 14, 6: 202 (1993). Type: Tanzania, Lushoto District: Usambara Mts, Nderema, *Holst* 2253 (B!, ?lecto.; K!, iso.); protolog has list of syntypes, *Holst* 2253 (B!, syn.), Gonja, 4236 (B!, syn.), Gonja, 4269a (B!, P!, syn.)

TANZANIA. Lushoto District: Amani, Oct. 1955, *Tanner* 2248! & Bomole Hill, June 1950, *Verdcourt* 274! & idem, Mar. 1998, *Luke* 5249!
DISTR. **T** 3; Mozambique, Zimbabwe
HAB. Moist forest; 600–1050 m
CONSERVATION NOTES. Widespread; least concern (LC)

SYN. *A. protensum* Schrad. var. *pseudohorridum* Hieron. in P.O.A. C: 82 (1895)
    A. pseudohorridum (Hieron.) Hieron. in E.J. 46: 362 (1911); Johns, Pterid. trop. East Africa checklist: 66 (1991)

NOTE. The typical species is restricted to Mauritius.
    This taxon resembles *A. protensum*, but that is always proliferous; and *A. friesiorum*, but that always has a creeping rhizome.

23. **Asplenium friesiorum** *C.Chr.* in N.B.G.B. 9: 181 (1924); Alston, Ferns W.T.A.: 57 (1959); Schelpe, F.Z. Pteridophyta: 178 (1970); Burrows, S. Afr. Ferns: 232, map, figs. (1990); Johns, Pterid. trop. East Africa checklist: 63 (1991); Faden in U.K.W.F. ed. 2: 28, t. 172 (1994). Type: Kenya, West Mt Kenya Forest Station, *Fries & Fries* 573 (B!, holo., BM, K!, S, UPS, iso.)

Low-epiphytic or less often terrestrial fern, occasionally in moist rock crevices; rhizome widely creeping, branching, up to 8 mm diameter, with fronds 3–4 cm apart and with iridescent brown narrowly triangular entire attenuate rhizome scales 3–5 mm long ending in a short hair-point. Fronds widely spaced, erect, not proliferous, 60–200 cm long with pinnae, thinly coriaceous. Stipe dark brown to almost black, (11–)15–70(–77?) cm long, with few brown lanceolate to ovate small scales similar to those on the rhizome. Lamina very narrowly elliptic to lanceolate, 26–101 × 10–26 cm, 1-pinnate to 2-pinnatifid, acuminate, basal pinnae somewhat reduced, apical segment deeply pinnatifid or less often the apical pinnae decrescent. Pinnae 14–35 pairs, alternate or subopposite, (shiny) dark green above, narrowly lanceolate, 6–14(–25) × 1–2.3(–3) cm, mostly shortly petiolate, unequally cuneate at the base tending to be slightly auriculate acroscopically, coarsely serrate to shallowly lobed with serrate lobes to ²/₃ of the way to the costa, the lobes at a 45° angle, apex linear-attenuate, glabrous on both surfaces except for scattered pale brown ovate-acuminate scales ± 1 mm long; costa closer to basiscopic margin, veins forked and at very sharp angle with the costa. Rachis dull brown to purple-brown to black, not winged, glabrous at maturity or nearly so. Sori usually many, oblong, 3–9 mm long, 1–3 mm wide at maturity, borne in 2 rows closely set along and parallel to the costa; indusium linear, membranous, entire, ± 1 mm wide, occasionally hidden when the sori are almost contiguous and sporangia are at full development. Fig. 4: 6–7, p. 25.

UGANDA. Ankole District: Rugongo, Nyagoma R., Jan. 1971, *Rwaburindore* 506!; Kigezi District: Kachwekano Farm, Sep. 1949, *Purseglove* 3113!; Masaka District: Lake Nabugabo, Aug. 1935, *Chandler* 1407!

KENYA. Northern Frontier District: Ndoto Mts, Manmanet ridge, Oct. 1995, *Bytebier & Kirika* 32!; Kericho District: SW Mau Forest, Kiptiget R. 16 km SSE of Kericho, June 1962, *Faden & Grumbley* 72/341!; Masai District: Chyulu Hills, Main forest N, Dec. 1993, *Luke & Luke* 3902!

TANZANIA. Bukoba District: Minziro Forest Reserve, July 2000, *Bidgood et al.* 4867!; Lushoto District: Magamba Kosti, Sep. 1981, *Mtui & Sigara* 69!; Iringa District: Ihangana Forest Reserve, Feb. 1962, *Polhill & Paulo* 1494!

DISTR. U 2, 4; **K** 1, 3–7; **T** 1–7; Congo-Kinshasa, Rwanda, Burundi, Zambia, Malawi, Mozambique, Zimbabwe, South Africa

HAB. Moist forest, bamboo forest, swamp forest, riverine thicket; may be locally common; 1100–2750(–3000) m

CONSERVATION NOTES. Widespread; least concern (LC)

SYN. *A. serra* Langsd. & Fisch. var. *natalensis* Baker, Syn. Fil., ed. 2: 485 (1883). Type: South Africa, Natal, *Buchanan* s.n. (K!, holo.)

A. monilisorum Domin in Preslia, 8: 7 (1929). Type: several syntypes including Kenya, "Karuris, E Kinobep", Feb. 1905, Imp. For. Herb. s.n. (K!, syn.); Tanzania, Kilimanjaro, *Volkens* 1272 (ubi?, syn., not found in B)

A. pseudoserra Domin in Preslia, 8: 6 (1929). Type: Tanzania, mainland W of Zanzibar, probably Nguru Mts, comm. Mar. 1885, *Last* s.n. (K!, iso.)

Tarachia friesiorum (C. Chr.) Momose in Journ. Jap. Bot. 35: 321, fig. 33–34 (1960)

NOTE. Looks slightly like *A. protensum*, but that has thinly pubescent proliferous fronds.

24. **Asplenium unilaterale** *Lam.*, Encycl. Méth. Bot. 2: 305 (1786); Sim, Ferns S. Afr. ed. 2: 152, t. 54 (1915); Tardieu in Mém. I.F.A.N. 28: 191, t. 37/1–2 (1953); Alston, Ferns W.T.A.: 56 (1959); Tardieu, Fl. Camér. 3, Ptérid.: 195, t. 29/1–2 (1964); Schelpe, F.Z. Pteridophyta: 174 (1970); Burrows, S. Afr. Ferns: 222, map, figs. (1990); Johns, Pterid. trop. East Africa checklist: 68 (1991); Faden in U.K.W.F. ed. 2: 28, t. 172 (1994). Type: Mauritius, no locality, *Commerson* s.n. (P!, holo.)

Usually on rocks but occasionally terrestrial or low epiphyte; rhizome long-creeping, 2–3 mm diameter, with dark brown clathrate lanceolate entire rhizome scales up to 2 mm long. Fronds spaced, erect. Stipe shiny dark brown to blackish, 7.5–22 cm long, glabrous. Lamina 17–30 × 5–10 cm, 1-pinnate, narrowly oblong in outline with lowest pinnae hardly reduced but often more lobed, apex decrescent end attenuate. Pinnae in 12–20 pairs, herbaceous, like a curved paralellogram or narrowly triangular, up to 3.5–6 × 0.5–1.2 cm, dimidiate for about $\frac{1}{2}$ of their length, petiolate, basiscopic base absent (= costa), acroscopic base parallel to rachis, serrate to double-serrate to dentate on the acroscopic margin, often doubly serrate towards the apex, basiscopic margin entire and formed by the costa for half or $\frac{2}{3}$, the upper half or $\frac{1}{3}$ double serrate or even slightly lobed; glabrous on both surfaces. Rachis shiny dark brown, glabrous. Sori 2–10(–14), confined to the outer $\frac{1}{4}$ to $\frac{1}{2}$ of the pinna, oblong to very narrowly oblong, set at an angle of about 30° to the costa, 2.5–3 mm long; indusium narrowly oblong, membranous, entire, 0.8 mm wide. Fig. 4: 8–9, p. 25.

UGANDA. Toro District: Kibale National Park, Dura R. bank, June 1997, *Poulsen & Nkuutu* 1270!; Kigezi District: Ishasha R., 7 km SW of Kirima, Sep. 1969, *Faden et al.* 69/1220!; Mengo District: Mabira Forest, 1904, *Dawe* 166!

KENYA. Kiambu District: Kamiti, Dec. 1969, *Faden & Evans* 69/2076!; North Kavirondo District: Kakamega Forest NE of Forest Station, Nov. 1969, *Faden et al.* 69/1989!; Teita District: Taita Hills, Bura R., Feb. 1955, *van Someren* 850!

TANZANIA. Kilimanjaro, Himo R., Marangu, Sep. 1964, *Beesley* 22!; Lushoto District: Derema, Nov. 1986, *Borhidi & Steiner* 86/425!; Iringa District: Udzungwa Mountain National Park, 'camp 222–pt 227', Sep. 2001, *Luke et al.* 7964!

DISTR. **U** 2–4; **K** 4, 5, 7; **T** 2, 3, 6, 7; tropical Africa; Malawi, Zimbabwe; Mascarene Is., S and E Asia and the Pacific Is.

HAB. In moist or riverine forest, always near streams or waterfalls where it may be locally common; 450–1950 m

CONSERVATION NOTES. Widespread; least concern (LC)

NOTE. *Luke & Luke* 5115 from **T** 7, Udzungwa Mts, has the sori spread over more of the pinna then usual.

25. **Asplenium obscurum** *Bl.*, Enum. Pl. Jav.: 181 (1828); Tardieu, Fl. Madag. Polypod. 1: 190, t. 27 fig. 1–3 (1958); Schelpe, F.Z. Pteridophyta: 174, t. 53g (1970); Burrows, S. Afr. Ferns: 221, map, figs. (1990); Johns, Pterid. trop. East Africa checklist: 66 (1991). Type: Indonesia, Java, Burangrang Mts, no collector indicated, may be a Blume specimen

Terrestrial; rhizome creeping, 4–5 mm diameter, with rather sparse brown narrowly triangular subentire acute clathrate rhizome scales ± 2 × 1 mm. Frond spaced, erect, not proliferous. Stipe matt green or brown, often fleshy, 12–35 cm long, with very few short hairs when young, glabrous with age and with a few scales similar to the rhizome scales near the base. Lamina oblong to narrowly oblong-acuminate, 18–35 × 8–17 cm, 1-pinnate, basal pinnae not reduced; pinnae in 15–20 pairs, membranous, petiolate, rhombic, falcate, 3.5–8.5 × 1.3–1.7 cm, costa forming the basiscopic margin for ± $^1/_3$ the length of the pinna, very unequally cuneate at the base, the costa forming the basiscopic margin at base, the acroscopic base parallel to rachis, serrate to mostly doubly serrate on the acroscopic margin and upper part of the basiscopic margin, broadly to narrowly acute, glabrous on both surfaces; rachis green or matt-brown with occasional multicellular hairs. Sori 11–30, narrowly oblong, set at about 45° to the costa about midway between the costa and margin, 6–7 mm long; indusium narrowly oblong, membranous, entire, 1 mm wide. Fig. 4: 10–11, p. 25.

TANZANIA. Morogoro District: NE Uluguru Mts, Kinole, Sep. 1970, *Faden et al.* 70/680! & Chita Forest reserve, Oct. 1984, *D.W. Thomas* 3955!; Iringa District: Mufindi, Kigogo R., Nov. 1985, *Schippers* T1141!
DISTR. **T** 6, 7; Malawi, Mozambique, Zimbabwe; Madagascar; Asia
HAB. Terrestrial in stream beds in deep shade in forest; 900–1650 m
CONSERVATION NOTES. Widespread; least concern (LC)

SYN. *A. serraeforme* Mett. in Abh. Senckenb. Nat. Ges. 3: 163 (1859) reimpr. in Mett. Farngatt. 6: 119, n. 75. t. 4 fig. 1 (1859). Type of unknown origin
Hemiasplenium obscurum (Bl.) Tagawa in Act. Phytotax. Geobot. 7: 83 (1938)

26. **Asplenium erectum** *Willd.*, Sp. Pl., ed. 4, 5: 328 (1810); Schelpe, F.Z. Pteridophyta: 175 (1970); Burrows, S. Afr. Ferns: 223, map, figs. (1990); Johns, Pterid. trop. East Africa checklist: 63 (1991); Faden in U.K.W.F. ed. 2: 28, t. 172 (1994). Type: Réunion, *Bory de St. Vincent* s.n. (B-W 19906, lecto., chosen by ?)

Terrestrial or low-level epiphyte, occasionally on rocks; rhizome 3–4 mm diameter, erect, with subulate dark brown clathrate attenuate entire rhizome scales 1.5–3 mm long. Fronds tufted, 15–50 cm long. Stipe matt grey-brown to black, 2–12 cm long, with narrow green wings in the upper $^1/_2$, with sparse scales similar to the rhizome scales when young, becoming glabrous with age. Lamina light green or dark green, erect or arching, narrowly elliptic in outline, 12–34 × 1.8–4.5 cm, 1-pinnate (sometimes to 2-pinnatifid on some lower pinnae), decrescent towards apex, not proliferous, firmly membranous. Pinnae up to 40 pairs, asymmetrically oblong (becoming flabellate towards the base of the frond), 0.8–2.5(–3.5) × 0.3–1 cm, shortly petiolate, ± auriculate acroscopically at the base with an acroscopic basal lobe separated to a greater or lesser degree, serrate except for the obtusely unequally cuneate base, rounded to acute, glabrous on both surfaces. Rachis dark matt-grey to black with 2 narrow green wings, glabrous. Sori up to 15 per pinna, narrowly oblong, 1.5–5 mm long, mostly set at an acute angle on both sides of the costa; indusium membranous, narrowly oblong, entire. Fig. 5: 1–2, p. 30.

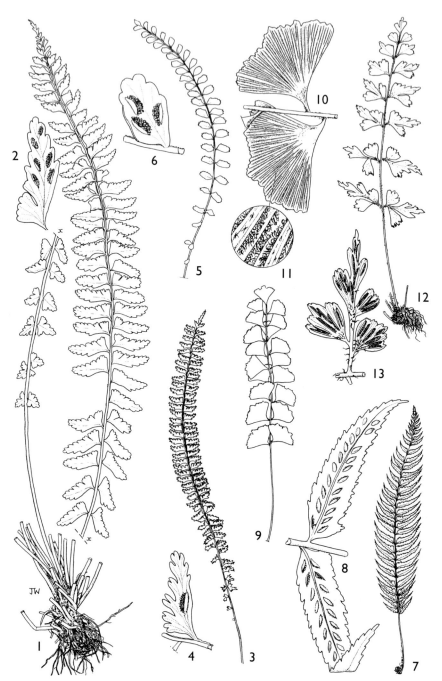

FIG. 5. *ASPLENIUM ERECTUM* — **1**, habit × ²⁄₃; **2**, pinna with sori × 2. *ASPLENIUM FORMOSUM* — **3**, habit, frond, not to scale; **4**, pinna × 2. *ASPLENIUM TRICHOMANES* — **5**, habit, frond × ²⁄₃; **6**, pinna with sori × 5. *ASPLENIUM SMEDSII* — **7**, habit × ¹⁄₁₂; **8**, pinnae × ²⁄₃. *ASPLENIUM LAURENTII* — **9**, habit, frond × ¹⁄₆; **10**, pinnae × ²⁄₃; **11**, detail of sori. *ASPLENIUM STUHLMANNII* — **12**, habit × ¹⁄₃; **13**, pinna with sori × 1. 1, 2, from *Bytebier & Kirika* 33; 3, 4, from *Luke* 8783; 5, 6, from *S.Paulo* 1011; 7, 8, from *Luke et al.* 8908; 9–11, from *Eggeling* 1474; 12, 13, from *Braithwaite* 316. Drawn by Juliet Williamson

UGANDA. Ruwenzori Mts, Mubuku valley above Nyabitaba Hut, Jan. 1967, *C.E. Smith* 4605!; Kigezi District: S Maramagambo Central Forest Reserve, 10 km up Kaizi–Bitereko road, Sep. 1969, *Faden et al.* 69/1107!; Mengo District: Mpanga Forest, 32 km W of Kampala, Feb. 1965, *Tweedie* 3003!

KENYA. Northern Frontier District: Ndoto Mts, Manmanet ridge, Oct. 1995, *Bytebier & Kirika* 33!; Nairobi, Karura Forest along Rui Raka R., Dec. 1969, *Faden & Evans* 69/2082!; Masai District: Chyulu Hills, Main Forest North, Dec. 1993, *Luke & Luke* 3877!

TANZANIA. Arusha District: Loitong, Ngurdoto Crater, Oct. 1965, *Greenway & Kanuri* 12108!; Ufipa District: Mbizi Forest Reserve, Oct. 1987, *Ruffo & Kisena* 2789!; Morogoro District: Nguru Mts near Maskati Mission, Mabega Mt, June 1978, *Thulin & Mhoro* 3069!

DISTR. **U** 2, 4; **K** 1, 3–7; **T** 2–4, 6, 7; tropical Africa from Guinea to Cameroon, southern Africa, South Africa; Madagascar, Mascarene Is.

HAB. Shaded forest floors; (?400 m in Shimba Hills) 900–2750 m (?–3650 m on Ruwenzori)

CONSERVATION NOTES. Widespread; least concern (LC)

SYN. *A. lunulatum* Sw. var. *erectum* (Willd.) Sim, Ferns S. Afr. ed. 2: 145 (1915)
  A. sphenolobium Zenker var. *usambarense* Hieron. in Z.A.E. 2: 14 (1911); Pic. Serm. in Webbia 32: 80 (1977) & in B.J.B.B. 55: 133 (1985). Type: Uganda, Butahu [Butagu] Valley, *Mildbraed* 2713 (B!, holo.)
  A. usambarense (Hieron.) Hieron. in Hedwigia, 60: 227 (1918); Johns, Pterid. trop. East Africa checklist: 68 (1991), *nom. illegit.*
  A. quintasii Gandog. in Bull. Soc Bot. Fr. 66: 305 (1919); Alston, Ferns W.T.A.: 57 (1959); Johns, Pterid. trop. East Africa checklist: 66 (1991), as '*quintarii*'. Type: Sao Tomé, *Quintas* 1342 (P!, holo.)
  A. erectum Willd. var. *usambarense* (Hieron.) Schelpe in Bol. Soc. Brot., Sér. 2, 41: 207 (1967); Schelpe, F.Z. Pteridophyta: 176, t. 53f (1970); Burrows, S. Afr. Ferns: 224, map, figs. (1990); Johns, Pterid. trop. East Africa checklist: 63 (1991); Schippers in Fern Gaz. 14, 6: 200 (1993), **syn. nov.**
  A. lunulatum sensu Johns, Pterid. trop. East Africa checklist: 65 (1991), *non* Sw.

NOTE. I am completely unconvinced about the difference between the varieties *erectum* and *usambarense*. These were generally separated as follows:

Lamina very narrowly oblong; basal auricle of lower pinnae not free;
  costa often whitish above; sori over 2 mm from the margin . . . . var. *erectum*
Lamina linear; basal auricle of the lower pinnae free; costa dark above . . . var. *usambarense*

but the free basal auricles are rare (less than 10% of the material available) and specimens overlap geographically completely with the main taxon; moreover, a single population (e.g. fide *Grimshaw* 93/1153 but observed on several multi-specimen collections) and occasionally even a single plant may show both free and united auricles. For our region at least I don't believe these varieties work – even though the type of *usambarense* comes from Uganda (!).

Much of the East African material has been named as *A. lunulatum* from the 1930s onwards; but that taxon is restricted to Southern Africa.

H.D. van Someren 116 from Kenya, Shimba Hills is most likely this taxon, even though the rhizome is missing. Van Someren did not indicate an altitude, but it must be below 450 m.

Several specimens have low numbers of sori per pinna and could key out to *A. normale*; however, the almost free acroscopic lobe of *A. erectum* is quite different from the pronounced but continuous acroscopic lobe of *A. normale*. These specimens are *Zerny* 175 and *Allen* 3734. Other, poor specimens may be difficult to key out.

27. **Asplenium formosum** *Willd.*, Sp. Pl., ed. 4, 5: 329 (1810); Tardieu in Mém. I.F.A.N. 28: 183, t. 37/3–4 (1953); Alston, Ferns W.T.A.: 57 (1959); Tardieu, Fl. Cameroun Pterid.: 196, t. 29/3–4 (1964); Schelpe, F.Z. Pteridophyta: 179 (1970); Burrows, S. Afr. Ferns: 230, map, figs. (1990); Johns, Pterid. trop. East Africa checklist: 63 (1991); Faden in U.K.W.F. ed. 2: 28, t. 172 (1994). Type: Venezuela, Caracas, *Bredemeyer* in herb. *Willdenow* 19908 (B-W, holo.)

On rocks or terrestrial; rhizome erect, 3 mm diameter, with brown lanceolate attenuate rhizome scales up to 3 mm long with a dark central stripe and pale entire borders. Frond tufted, erect or slightly arching, not proliferous. Stipe shining dark brown to black, 0.5–6 cm long, with 2 narrow pale brown wings. Lamina very narrowly oblong, 10–31 × 1.5–3.5 cm, 1-pinnate to 2-pinnatifid, with the lower pinnae gradually reduced, apical pinnae decrescent but terminal pinna similar to others, though with 2 incised margins. Pinnae (15–)25–50 pairs, opposite or alternate, chartaceous, 0.9–1.8 × 0.4–0.8 cm, shortly petiolate, unequally narrowly oblong-rhombic, deeply incised into linear obtuse segments ± 1 mm wide, the acroscopic basal lobe parallel to or sometimes overlapping the rachis, the basiscopic margin entire, the acroscopic margin incised (see above), both surfaces glabrous; costa running close to basiscopic margin. Rachis dark brown to black with a narrow pale wing, glabrous. Sori 1–2(–3) per pinna, ovate to elliptic, all facing acroscopically, 2–4 mm long; indusium narrow-elliptic, membranous, entire, 0.7–1.3 mm wide. Fig. 5: 3–4, p. 30.

UGANDA. Karamoja District: Mt Debasien, July 1949, *van Someren* 615!
KENYA. North Kavirondo District: Kakamega Forest along Isiukhu R., Dec. 1969, *Faden & Rathbun* 69/2113!
TANZANIA. Arusha District: Lake Duluti, 11 km E of Arusha, Sep. 1965, *Beesley* 166!; Kigoma District: Mt Livandabe, May 1997, *Bidgood et al.* 4202!; Iringa District: Udzungwa Mountain National Park, 'below Pt. 245', June 2002, *Luke & Luke* 8783!
DISTR. U 1; K 5; T 2, 4, 7; tropical Africa including Cameroon to Sudan, S to Zambia and Mozambique; Madagascar; India; Sri Lanka; tropical America
HAB. Moist to dry forest, riverine forest; on moist rock or terrestrial; 600–1800 m
CONSERVATION NOTES. Widespread; least concern (LC)

NOTE. A specimen from T 2, S of Mt Meru, *Schippers* T1198! has unusual low numbers of pinnae, as few as 15 pairs on fertile fronds.

28. **Asplenium trichomanes** *L.*, Sp. Pl. 2: 1080 (1753); Sim, Ferns S. Afr. ed. 2: 140 44 fig. 1 (1915); Schelpe, F.Z. Pteridophyta: 174 (1970); Burrows, S. Afr. Ferns: 228, map, figs. (1990); Johns, Pterid. trop. East Africa checklist: 68 (1991); Thulin, Fl. Somal. 1: 13 (1993); Faden in U.K.W.F. ed. 2: 28 (1994). Type: Europe, LINN 1250/12 (LINN, lecto.)

Rhizome erect, with clathrate dark brown subulate attenuate rhizome scales up to 4 × 0.5 mm, often with a dark central stripe and paler narrow borders. Fronds tufted, erect to arching, firmly membranous, not proliferous. Stipe dark brown, 1–9 cm long, usually less than ¹/₆ the length of the lamina, glabrous. Lamina narrowly linear in outline, 6–23 × 1–1.5 cm, 1-pinnate, lower pinnae slightly reduced. Pinnae in 26–32 pairs, broadly oblong to obovate, up to 8 × 5.5 mm, shortly petiolate, crenate, glabrous on both surfaces or with a few minute scales beneath. Rachis dark brown, glabrous. Sori 1–6 per pinna, oblong at first, almost covering the dorsal surface of the pinna at maturity; indusium narrowly oblong, membranous, to 0.3 mm wide, entire. Fig. 5: 5–6, p. 30.

UGANDA. Kigezi District: Muhavura Mt, Dec. 1957, *Allen* 3733!
KENYA. Turkana District: Murua Nysigar, Sep. 1963, *Paulo* 1011! & idem, Dec. 1988, *Beentje, Powys & Dioli* 3943!
TANZANIA. Arusha District: Mt Meru, Engare Nanyuki R., June 1965, *Vesey-Fitzgerald* 4701!
DISTR. U 2; K 2; T 2; Zimbabwe, South Africa; NE tropical Africa; Europe, Australasia, Americas
HAB. Low epiphyte in rocky forest; 1750–2700 m
CONSERVATION NOTES. Widespread; least concern (LC)

29. **Asplenium smedsii** *Pic.Serm.* in Webbia 32(1): 60, t. 2–3 (1977); Johns, Pterid. trop. East Africa checklist: 67 (1991); Faden in U.K.W.F. ed. 2: 28, t. 172 (1994). Type: Ethiopia, Shashamane District: road to Adaba, *de Joncheere* ETS 50 (herb. Pic. Serm. 26870)

Epiphyte, lithophyte or terrestrial; rhizome short, erect, with chestnut-brown triangular long-attenuate rhizome scales with entire or fimbriate-lobed margin, ending in hair-tip and (5–)7–11 × 2.5–4 mm. Fronds tufted, erect, olive-green, not proliferous. Stipe matt brown, sometimes green beneath, 8–30 cm long, with dense scales near base, otherwise with scattered narrowly triangular scales to 10 × 3 mm. Lamina papyraceous, elliptic to oblong-elliptic in outline, 18–47 × 7–15(–24) cm, 1-pinnate; lower pinnae slightly smaller, apical pinnae gradually decrescent. Pinnae in 14–24 pairs, equidistant, stalked, triangular or lanceolate-triangular, (2–)4–8(–12) × (0.7–)1.2–1.6 cm, base unequal, the acroscopic base broadly cuneate, the basiscopic base narrowly cuneate-excised, margin deeply serrate with often bifid teeth, apex long-attenuate, glabrous above, sparsely scaly beneath. Rachis brown, with two narrow green lines, with sparse scales. Sori many per pinna, at a 45° angle to the costa, elliptic, 2.5–6 mm long; indusium membranous, entire, 1–1.5 mm wide. Fig. 5: 7–8, p. 30.

UGANDA. Mt Elgon, Suam R., Sep. 1961, *E.J. Brown* EA 12494! & Mt Elgon, May 1997, *Wesche* 1340!; Masaka District: Bugala, Sese, July 1935, *A.S. Thomas* 1388!
KENYA. Meru District: Nyambeni Hills, Kirima peak, Oct. 1960, *Verdcourt & Polhill* 2955!; Trans Nzoia District: Mt Elgon by Makingeny cave, Feb. 1985, *Beentje* 1958!; Teita District: Taita hills, Ngangao, Feb. 1977, *Faden & Faden* 77/327!
TANZANIA. Kilimanjaro: between Umbwe and Weru Weru Rs., Aug. 1932, *Greenway* 3194!; Lushoto District: W Usambara, Mkusu valley between Mkuzi and Kifungilo, Apr. 1953, *Drummond & Hemsley* 2229!; Rungwe District: N slope of Mt Rungwe, Feb. 1961, *Richards* 14268!
DISTR. **U** 3, 4; **K** 3–5, 7; **T** 2, 3, 6 (see Note), 7; Sudan, Ethiopia, Congo-Kinshasa, Rwanda, Burundi
HAB. Moist upland forest; low epiphyte, lithophyte or terrestrial; 1200–2500 m
CONSERVATION NOTES. Widespread; least concern (LC)

SYN. *A.* sp. *B* of Faden in U.K.W.F. ed. 1: 65 (1974)

NOTE. Close to *A. boltonii* but not proliferous and with slightly different pinnae, as well as broader rhizome scales.
    Tanzania **T** 6, Morogoro District: Kanga Mountain, Dec. 1987, *Lovett & Thomas* 2760! is most likely this taxon but has more dissected pinnae, the individual pinnae being lobed halfway to the costa.
    Faden 72/306 and *Kerfoot* 3764 from **K** 5, and *Faden* 69/94 from Mt Kenya have much smaller pinnae (2–3 cm long) and smaller rhizome scales. I would have said that these might represent an allied but separate taxon, if other sheets of *Faden* 72/306 had not shown gradual variation to the normal form of *A. smedsii*.
    Pócs & Lundqvist 6475/G from **T** 6, S Uluguru above Kibungo mission, Oct. 1971, is probably this species but has pinnae to 15.5 × 2.9 cm; it lacks a rhizome.
    Faden et al. 71/891 from **K** 4, Mt Kenya, Castle Forest Station, Nov. 1971, is this species but has the acroscopic base produced into a lobe on many of the pinnae. I believe this is just a form.

30. **Asplenium laurentii** *Christ* in Bull. Herb. Boiss. 4: 663 (1896); Tardieu in Mém. I.F.A.N. 28: 174 (1953); Alston, Ferns W.T.A.: 57 (1959); Tardieu, Fl. Cameroun Pterid.: 185, t. 26/6–7 (1964); Johns, Pterid. trop. East Africa checklist: 65 (1991). Type: Congo-Kinshasa, lower Congo R., Vungu, *Laurent* s.n. (G? BR?)

Epiphyte; rhizome erect, ± 5 mm diameter, with dark brown subulate entire rhizome scales 8–13 mm long. Fronds tufted, not proliferous. Stipe blackish grey, 7–20 cm long, with sparse subulate dark brown scales. Lamina lanceolate, 17–30 × 5–9 cm, 1-pinnate. Pinnae in 5–11 pairs, (sub)opposite, fan-shaped, 3–4 × 4–6 cm, margin entire, apical margin irregularly serrate, petiolate, basal pinnae not reduced, terminal pinna similar to others; veins flabellate; surfaces with thread-like slightly lobed black scales. Sori many, along the veins, linear, 18–30 mm long, opening in both directions, not reaching the margin; indusium membranous, entire, ± 0.3 mm wide. Fig. 5: 9–11, p. 30.

UGANDA. Bunyoro District: Budongo forest, Dec. 1934, *Eggeling* 1474! & idem, Apr. 1971, *Synott* 564!

DISTR. **U** 2; Gabon, Congo-Kinshasa
HAB. Moist forest; no altitude given
CONSERVATION NOTES. Widespread; least concern (LC)

31. **Asplenium diplazisorum** *Hieron.* in E.J. 46: 351 (1911); Tardieu, Mém. I.F.A.N. 28: 179, t. 34/1 (1953); Alston, Ferns W.T.A.: 57 (1959); Tardieu, Fl. Cameroun Pterid.: 199 (1964); Johns, Pterid. trop. East Africa checklist: 62 (1991). Type: Cameroon, Lolo village [Lolodorf], *Staudt* 22 (B!, P!, syn.) & 194 (B!, P!, syn.)

Low epiphyte or terrestrial; rhizome erect, short, with deltoid rhizome scales ending in an attenuate tip [not visible on our specimen]. Fronds tufted, erect, not proliferous. Stipe 8–16 cm long, with sparse scales and near apex articulated hairs [in ours look like stalked glands]. Lamina herbaceous, oblong, 10–15 × 5–7 cm, 1-pinnate, basal pinnae not or hardly reduced, terminal pinna deltoid-oblong, dentate; pinnae in 6–8 pairs, alternate, petiolulate, trapezoid-oblong, to 4 × 1.7 cm, acroscopic base truncate-cuneate and parallel to the rachis, basiscopic margin obliquely cuneate, margin crenate-dentate, apex rounded or acute, glabrous. Rachis narrowly winged [I think], with sparse ?stalked glands. Sori many, on the veins at 45° angle to costa, not reaching costa or margin, linear, 3–8 mm long; indusium membranous, entire, ?0.3 mm wide.

UGANDA. Mengo District: Semunya forest near Entebbe, Feb. 1950, *Dawkins* 524!
DISTR. **U** 4; W Africa from Sierra Leone to Congo-Kinshasa
HAB. Swamp forest, where locally common, on stumps or mounds; ± 1140 m
CONSERVATION NOTES. Widespread; least concern (LC)

32. **Asplenium stuhlmannii** *Hieron.*, P.O.A. C: 83 (1895); Alston, Ferns W.T.A.: 59 (1959); Pic. Serm. in B.J.B.B. 55: 152 (1985); Johns, Pterid. trop. East Africa checklist: 67 (1991); Faden in U.K.W.F. ed. 2: 30, t. 173 (1994). Lectotype: Country unclear, Kanesse, *Stuhlmann* 936 (B!, lecto., chosen here)

Epiphyte or lithophyte; rhizome short-creeping or erect, to 5 mm diameter, with dark brown narrowly triangular rhizome scales 2.5–8 × 0.3–0.5 mm, attenuate at apex, margin paler and entire or with outgrowths. Fronds closely spaced or tufted, to 40 cm long, erect, not proliferous. Stipe dark brown, 6–28 cm long, at first densely set with scales similar to those of the rhizome, later becoming subglabrous except near the base. Lamina narrowly oblong, 6–38 × 1.6–5(–8.8) cm, 1-pinnate to 2-pinnatifid or 2-pinnatisect (rarely 3-pinnatifid), the lower pinnae slightly reduced, with a terminal pinna similar to lateral ones; pinnae in 8–15 pairs, opposite in mid-frond, narrowly ovate-triangular to ovate in outline, up to 3(–4.5) × 1.6(–2.5) cm, subsessile, palmately to pinnately divided into 3–5 serrate lobes to 1 cm long, the central lobe prominent and serrate (rarely to 3.7 × 1.2 cm, and deeply pinnatisect), with a few minute scales or glabrous. Rachis dark brown at first densely set with shining scales similar to those on the stipe, glabrescent. Sori many, may cover lower pinna surface, linear, (1–)3–10 mm long; indusium linear, membranous, 0.4–0.5 mm wide, entire to slightly erose. Fig. 5: 12–13, p. 30.

UGANDA. Ankole District: Nyakatokye–Bunyenye, Kashari, Mar. 1986, *Rwaburindore* 2235!
KENYA. Kitui District: Galunka, May 1902, *Kassner* 793!; Teita District: Msau–Mbololo road, near Mwambirwa Forest Station, Sep. 1970, *Faden et al.* 70/520!
TANZANIA. Bukoba District: Ilomera Mission, Sep. 1974, *Balslev* 11!; Lushoto District: Lushoto, June 1961, *Braithwaite* 316!; Iringa District: Ruaha National Park, Magangwe Hill, May 1972, *Bjørnstad* 1799! & 1800!
DISTR. **U** 2; **K** 3?, 4, 7; **T** 1, 3, 7; Burundi, Zambia, Mozambique, Zimbabwe
HAB. Moist forest, miombo woodland, sometimes (often?) on rock outcrops and in rock crevices; 1200–1700 m but type presumably from much lower altitude
CONSERVATION NOTES. Widespread; least concern (LC)

SYN. *A. stuhlmannii* Hieron. var. *laciniata* Hieron., P.O.A. C: 83 (1895). Type: Sudan, Niamniam land, on Gumango (R.), *Schweinfurth* 3915 (B!, holo.)
    A. ramlowii Hieron. in E.J. 46: 372 (1911); Schelpe, F.Z. Pteridophyta: 180 (1970); Burrows, S. Afr. Ferns: 245, map, figs. (1990). Type: Tanzania, Tanga, *Ramlow* 16 (B!, holo.), **syn. nov.**

NOTE. When one compares the types and protologues of *A. stuhlmannii* and *A. ramlowii* hardly any differences are apparent. The size range of rhizome scales in *ramlowii* is slightly greater, and pinnae number is slightly smaller in *ramlowii*. The type of *A. ramlowii* is a minimalist bunch of scraps, but probably really the same as the rest of the taxon so named.
    Var. *laciniatum* looks quite different, much more incised, but *Uhlig* 106 has a single frond divided in the same way as the type of *laciniatum*, while the other fronds are as in the typical variety. The dimensions are indicated in brackets in the above description.
    The pinnae look quite like uppermost few pinnae of *A. albersii* (now in *A. aethiopicum*).
    The Kenyan *Kassner* specimen and **T** 1, Moru Kopjes, Apr. 1961, *Greenway & Miles Turner* 10033! are the only ones that have the pinnae divided into discrete pinnules, which in turn are lobed. Another Kenyan specimen, **K** 3, Uasin Gishu District: Kipkarren, Aug. 1931, *Brodhurst-Hill* 49! is sterile, but looks very much like this species.
    Information received after the text was completed: Dr A Braithwaite informs me re *Braithwaite* 316, that there is a clear cytological difference between this material and *A. stuhlmannii*. It has the same chromosome number as material of *A. simii* Dr Braithwaite collected from Knysna, South Africa. i. e. n = 144, while *A. stuhlmannii* is tetraploid with n = 72. *A. simii* is recorded as 'excluded species' (page 69) but this will obviously have to be revisited during follow-up treatments!

33. **Asplenium auritum** *Sw.* in Schrad., Journ. Bot. 1800, 2: 52 (1800); Schelpe, F.Z. Pteridophyta: 177 (1970); Burrows, S. Afr. Ferns: 244, map, figs. (1990); Johns, Pterid. trop. East Africa checklist: 62 (1991). Type: Jamaica, *Swartz* s.n. (S, lecto.; UPS, iso.)

Lithophyte or epiphyte; rhizome erect or suberect, up to 10 mm diameter, with shining dark brown entire attenuate rhizome scales up to 6 mm long. Frond tufted, arching, not proliferous, thinly coriaceous, the fertile fronds twice as long as the sterile fronds. Stipe matt grey to greyish green, often darker towards the base, 5–22 cm long, glabrous. Lamina triangular-lanceolate to narrowly oblong-lanceolate, 1-pinnate to shallowly 2-pinnatifid, sometimes on lowermost pinnae the acroscopic lobe almost free, 8–26 × 3.2–10 cm, basal pinnae not reduced, decrescent upwards into a deeply pinnatifid apex. Pinnae up to 15 pairs, narrowly oblong or lanceolate, coriaceous, 1.8–5.5 × 0.4–2 cm, petiolate; on sterile fonds pinnae smaller and crenate, on larger fertile fronds pinnae base asymmetrically cuneate with a free acroscopic lobe or auricle, margin deeply pinnatifid or pinnate into cuneate-oblong irregularly crenate-serrate segments; apex obtuse to acute. Rachis matt grey-green, glabrous except for scattered black hair-like scales ± 1 mm long. Sori several and one per pinna lobe (several in basal acroscopic lobe) or less often few in distal part of pinna, along vein at angle of ± 30°, oval, 2–4 × 1.5 mm, oblong; indusium linear, membranous, entire, 0.6–0.8 mm wide.

UGANDA. Bunyoro District: Budongo, July 1953, *H.D. van Someren* 751!; Kigezi District: Ishasha Gorge, Apr. 1998, *Hafashimana* 511!; Elgon, Mbale, Dec. 1950, *H.D. van Someren* 570!
TANZANIA. Lushoto District: E Usambara, N'dola–Nguo track, Feb. 1954, *Faulkner* 1353!; Morogoro District: NE Uluguru Mts, Kinole, Sep. 1970, *Faden et al.* 70/684!; Iringa District: Udzungwe Mts, Sanje Falls, Nov. 1995, *de Boer et al.* 738!
DISTR. **U** 2, 3; **T** 3, 6, 7; Congo-Kinshasha, Malawi, Mozambique, Zimbabwe; Madagascar; Mascarenes; tropical America
HAB. On moss-covered rock or epiphytic in moist forest; 900–1450 m
CONSERVATION NOTES. Widespread; least concern (LC)

34. **Asplenium ceii** *Pic.Serm.* in Nuov. Giorn. Bot. Ital. ser. 2. 47: 11, t. 3 (1940); Pic. Serm. in B.J.B.B. 55: 128 (1985); Johns, Pterid. trop. East Africa checklist: 62 (1991); Faden in U.K.W.F. ed. 2: 28, t. 172 (1994). Type: Ethiopia, Kaffa, between Bonga and Uota, *Cei* 14 (FT, holo.)

Epiphytic; rhizome erect, ± 1 cm diameter, covered with old stipe bases with greyish-brown, lanceolate-subulate, subentire scales up to 10 × 1.5 mm, composed of clear thin-walled cells, attenuate into a hair-like tip. Frond tufted, erect, not proliferous. Stipe matt-grey-green to brown, 5–30 cm long and 3 mm diameter, subglabrous, with a few scales towards the base similar to the rhizome scales. Lamina dark shiny green, herbaceous to coriaceous, 18–40 × 10–19 cm, oblong, erect, 1-pinnate; basal pinnae not reduced, terminal pinna similar to lateral ones. Pinnae in 2–9 sub-opposite pairs with a terminal pinna similar to the other pinnae, petiolulate, lanceolate to oblong, 8–14 × 1.8–2.5 cm, the terminal pinna 14–16 × 2.5–3.5 cm, acroscopic base cuneate and often parallel to the rachis, basiscopic base attenuate-cuneate, margin irregularly and shallowly undulate-crenate, attenuate to acuminate, glabrous; veins ending short of the margin. Rachis matt-purplish-brown, at first set with dark irregularly fimbriate scales up to 2 mm long, later becoming glabrous. Sori many, along the veins at 45° angle to costa, extending from near the costa to more than halfway to the margin, (4–)9–13 mm long; indusium membranous, entire, 0.5–0.8 mm wide. Fig. 6: 1–2, p. 37.

UGANDA. Toro District: Kibale Forest, Kanyawara, Sep. 1997, *Hafashimana* 332! & Kigezi District: Bwindi Forest, Ihihizo, Aug. 1998, *Hafashimana* 773!; Busoga District: Lolui Island, May 1965, *Jackson* 3/11/5/65
KENYA. Elgon, about 1919, *Gardner* 1028!; North Kavirondo District: Isiukhu R. along Kambiri–Vihiga road, Dec. 1969, *Faden & Rathbun* 69/2104! & Kakamega Forest, near forest station, Oct. 1981, *Gilbert & Mesfin* 6671!
TANZANIA. Kilosa District: Ukaguru Mts, Mt Kifigo, May 1972, *Pócs et al.* 6591A!; Iringa District: Udzungwa Mountain National park, camp 239, Oct. 2001, *Luke et al.* 8123!
DISTR. U 2, 3; K 3, 5; T 6, 7; Ethiopia, Burundi, Malawi, Mozambique, Zimbabwe
HAB. Moist forest; 1200–1700(–2400) m
CONSERVATION NOTES. Widespread; least concern (LC) – but only six collections from our area

SYN. *A. atroviride* Schelpe in Bol. Soc. Brot., Sér. 2, 41: 204 (1967); Schelpe, F.Z. Pteridophyta: 173 (1970); Burrows, S. Afr. Ferns: 218, map, figs. (1990); Johns, Pterid. trop. East Africa checklist: 62 (1991). Type: Zimbabwe, Vumba Mts., Witchwood, *Schelpe* 5446 (BOL, holo., BM!, iso.), **syn. nov.**

NOTE. *A. ceii* resembles a small *gemmiferum*, but occurs away from water, usually in deep shade. Christensen felt this might be a variety of *A. gemmiferum*, but Pichi Sermolli feels it is distinct enough. The Schelpe name was published, I believe, because the author was unaware of the existence of *A. ceii*.
   H.D. van Someren 929 collected in May–June 1957 in U 2, Kigezi District: Kayonza forest might be *A. ceii* but has altogether larger parts – lamina to 52 × 26 cm, pinnae to 17 × 3.5 cm, sori 12–23 mm.

35. **Asplenium anisophyllum** *Kunze* in Linnaea 10: 511 (1836); Schelpe, F.Z. Pteridophyta: 170 (1970); Burrows, S. Afr. Ferns: 214, map, figs. (1990); Johns, Pterid. trop. East Africa checklist: 61 (1991). Type: South Africa, between Umzimvubu [Omsamwubo] and Umsikaba (Omsamcaba] Rs., *Drège* s.n. (LZ†, holo.; B!, lecto., B!, BM! (fragm.), K! (fragm.) , iso.) lectotype chosen by Roux, 1986

Epiphyte or less often terrestrial; rhizome to 2 cm diameter, erect, with pale brown to brown concolorous lanceolate scales 6–12 × 1.5–3.5 mm, acute to acuminate, margin with thin lobes ('fimbriate'). Fronds tufted, erect or arching, not proliferous. Stipe matt-purplish-brown, 13–48 cm long, to 4 mm diameter, glabrous except for a few scales at the base similar to those on the rhizome. Lamina ovate to ovate-elliptic, 20–88 × 8–32 cm, 1-pinnate, with the basal pairs of pinnae somewhat reduced, terminal pinna large and similar to laterals, 5–12 × 1.6–2.5 cm. Rachis glabrous or with few very narrow scales to 1 mm. Pinnae in 7–20 pairs, narrowly lanceolate and a bit falcate, the largest 5–16 × 1.5–3 cm, apex acuminate or acute, base truncate and parallel to rachis on the acroscopic side, oblique on the basiscopic side, margin regularly crenate with a vein ending in each crenation, glabrous above but with a few

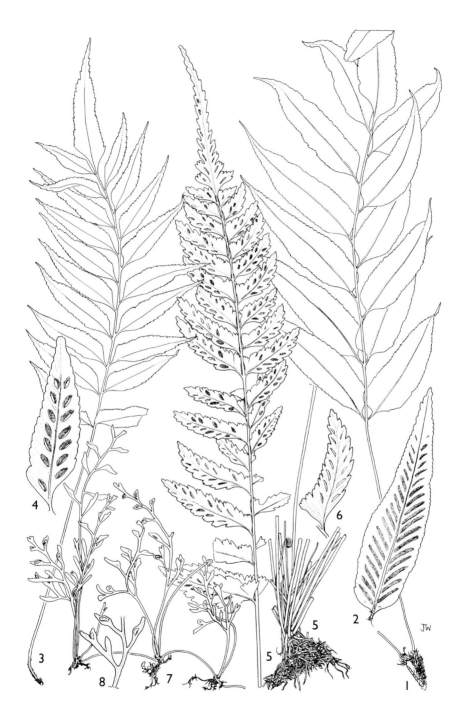

FIG. 6. *ASPLENIUM CEII* — **1**, habit, frond × ¹/₃; **2**, pinna × ²/₃. *ASPLENIUM ANISOPHYLLUM* — **3**, habit, frond × ¹/₃; **4**, pinna × ²/₃. *ASPLENIUM INAEQUILATERALE* — **5**, habit, frond × ²/₃; **6**, pinna with sori × 1. *ASPLENIUM MANNII* — **7**, habit × 1; **8**, pinnae × 1¹/₂. 1, from *Hafashimana* 773; 2, from *Jackson* 311565; 3, from *Faden* 69/1004; 4, from *Hafashimana* 424; 5, 6, from *Loveridge* 393; 7, 8, from *Purseglove* 2378. Drawn by Juliet Williamson.

scattered minute branched linear scales on the costae and veins below. Rachis glabrous. Sori many per pinna, at 45° to costa, extending from near the costa to $\frac{1}{2}$ way to the margin, 2.5–6 mm long; indusium membranous, to 1.2 mm wide, entire. Fig. 6: 3–4, p. 37.

UGANDA. Bunyoro District: Budongo Forest, Nyakafunjo Nature Reserve, Nov. 1996, *Hafashimana* 101!; Kigezi District: Ishasha Gorge, Nov. 1997, *Hafashimana* 424!; Mengo District: Mpanga Forest Reserve, 5 km E of Mpigi, Sep. 1969, *Faden et al.* 69/1004!
KENYA. Teita District: Mbololo Hill, Mraru ridge, July 1969, *Faden et al.* 69/826! & Kasigau, Bungule route, Nov. 1994, *Luke* 4184!
TANZANIA. Pare District: Shengena Forest Reserve, Gonja, Feb. 1988, *Kisena* 358!; Lushoto District: East Usambaras, May 1926, *Peter* 40244!; Njombe District: Lupembe, Apr. 1931, *Schlieben* 573!
DISTR. U 2, 4; K 7; T 3, 7; Angola, Malawi, Mozambique, Zimbabwe, South Africa; Madagascar
HAB. Epiphyte (low to high) or terrestrial in moist forest; 1000–1600 m
CONSERVATION NOTES. Widespread; least concern (LC)

SYN. *A. geppii* Carr., Cat. Afr. Pl. Welw. 2, 2: 269 (1901); F.W.T.A. Pterid.: 56 (1959). Type: Angola, Pungo Andongo, Mata de Pongo, *Welwitsch* 97 (BM, K, iso.)
  A. subauriculatum Hieron. in E.J. 46: 350 (1911); Pic. Serm. in B.J.B.B. 55: 152 (1985). Types: Cameroon, Lolodorf, *Staudt* 180 (B!, syn.; web image!) & Congo, Arthington falls, *Buttner* 194 (B!, syn.; web image!)

NOTE. Very similar to *A. elliottii* except for the terminal pinna in *A. elliottii* being proliferous. *A. boltonii* and *A. gemmiferum* are also close; but are both proliferous; and *A. ceii* has fewer pinnae which are much darker and glossy.
Note *A. anisophyllum* Kunze var. *aequilateralis* Hieron. = *A. elliottii*.

36. **Asplenium inaequilaterale** Willd., Sp. Pl. ed. 4, 5: 322 (1810); Tardieu in Mém. Inst. Fr. Afr. Noire, 28: t. 34/4–5 (1953); Alston, Ferns W.T.A.: 57 (1959); Schelpe, F.Z. Pteridophyta: 176, t. 53c (1970); Burrows, S. Afr. Ferns: 222, map, figs. (1990); Johns, Pterid. trop. East Africa checklist: 65 (1991); Schippers in Fern Gaz. 14, 6: 201 (1993); Faden in U.K.W.F. ed. 2: 29, t. 173 (1994). Type: Réunion, *Bory de St Vincent* in herb. *Willdenow* 19896 (B-W, holo.)

Terrestrial or on rocks, rarely a low epiphyte; rhizome erect, 4–8 mm diameter, with dark brown entire clathrate lanceolate attenuate rhizome scales 2–3 × 0.6 mm, the margin mid-brown. Frond tufted, erect, not proliferous. Stipe pale to dark matt-brown or greenish-brown, 4–19 cm long, sparsely covered at first with minute hair-like scales, rarely with a few stalked glands, glabrous at maturity. Lamina narrowly ovate, 12–32 × 4–13 cm, 1-pinnate, with lowest pinnae not markedly reduced; apical segment lobed, crenate, attenuate. Pinnae in 8–20 pairs, herbaceous, narrowly to very narrowly oblong or roughly lanceolate, sometimes falcate, 1.5–6.3 × 0.7–1.2 cm, unequally cuneate at the base, acroscopic lobe well-developed, crenate or doubly crenate, apex obtuse to attenuate, shortly petiolate; glabrous on both surfaces. Rachis pale brown, narrowly winged on the upper part of the lamina. Sori 3–16, linear (to oblong in small fronds), set at an angle of 30–45° to the costa, 2–7 mm long; indusium linear to oblong, membranous, entire, 0.5–1 mm wide. Fig. 6: 5–6, p. 37.

UGANDA. Bunyoro District: Rabongo forest, May 1993, *Sheil* 1662!; Mt Elgon, Sipi Bugishu, Dec. 1938, *A.S. Thomas* 2597!; Mengo District: 1.5 km NE of Nansagazi, Sep. 1969, *Faden et al.* 69/1029!
KENYA. Meru District: Lower Imenti Forest, Mar. 1970, *Faden & Evans* 70/123!; North Kavirondo District: Kakamega Forest along Kubiranga stream, Mar. 1977, *Faden & Faden* 77/828!; Teita District: Taita Hills, Mbololo, May 1985, *NMK Taita Hills Expedition* 409!
TANZANIA. Meru District: S slopes of Mt Meru, 1917, *Leighton* s.n.; Lushoto District: Baga Forest Reserve, May 1987, *Kisena* 631!; Iringa District: Udzungwa Mountain National Park, Mwaya–Mwanihana route, Nov. 1997, *Luke & Luke* 4890!
DISTR. U 1–4; K 3–5, 7; T 1–4, 6–8; tropical Africa and south to South Africa; Comoro Is., Réunion

HAB. Moist forest, on rocks by streams or on forest floor; 700–2100 m
CONSERVATION NOTES. Widespread; least concern (LC)

SYN. *A. brachyotus* Kunze in Linnaea, 10: 512 (1836); Johns, Pterid. trop. East Africa checklist: 62
(1991). Type: South Africa, between Umzimvubu [Omsamwubo] and Umsikaba
[Omsamcaba], *Drège* s.n. (not found; a K specimen without locality data has been chosen
as 'lectotype' by Roux)
A. erectum Willd. var. *brachyotus* (Kunze) Sim, Ferns S. Afr. ed. 2: 138, t. 66 (1892)
A. laetum sensu Sim, Ferns S. Afr. ed. 2: 150, t. 50 (1915); Johns, Pterid. trop. East Africa
checklist: 65 (1991), as *A. lactum*; *non* Sw. (1806)
A. laetum Sw. var. *brachyotus* (Kunze) Bonap., Not. Ptérid. 16: 60 (1925)
A. suppositum Hieron. in E.J. 46: 353 (1911); Alston, Ferns W.T.A.: 57 (1959); Johns, Pterid.
trop. East Africa checklist: 67 (1991); Faden in U.K.W.F. ed. 2: 29 (1994). Type: Angola,
Pungo Andongo, *Soyaux* 234 (B!, holo.; K!, iso.)

NOTE. This taxon differs from *A. unilaterale* chiefly in its erect rhizome and tufted fronds. It is also
close to *A. macrophlebium* which is usually proliferous, and has the veins mostly unbranched.
The mention of *Asplenium pulchellum* Hiern in P.O.A. C is probably based on this taxon.
Faden felt several specimens from Meru, which he named *A. suppositum*, differed in the
smaller, more finely toothed and more numerous pinnae, which give this plant a look
different from *A. inaequilaterale* and he considered them distinct; they grow side by side with
A. inaequilaterale in the lower Imenti forest near Meru. The type of *A. suppositum*, from
Angola, is certainly *A. inaequilaterale*.
A series of specimens from **U** 1, Mt Debasien, collected by *H.D. van Someren* (602, 603, 613,
614, 627) all key to this taxon but have a slightly different look to the pinnae, with 2-toothed
sub-lobes. *Richards* 14239 from **T** 7 is similar.
A specimen from Tanzania, Tanga District: Mtai Forest Reserve, *Lovett & Hamilton* 891! has
a different look and feel from *A. inaequilaterale*, to which it seems closest. It might otherwise
be close to *A. christii* from which it differs in obtuse pinna apices. It is a low epiphyte with
scales dark brown, clathrate, narrowly ovate to triangular, to 1.2 × 0.5 mm. Lamina oblong,
12 × 3–4.5 cm. Pinnae in 6 pairs, asymmetrically ovate, the largest 1.8–2.3 × 1.3–1.5 cm,
obtuse, margin crenualate, glabous or nearly so; terminal pinna similar to laterals. Sori 9–13
per lobe, reaching neither costa nor margin, on the branched veins and 3–5 mm long;
involucre whitish, membranous, entire, to 0.4 mm wide. Epiphytic on base of tree trunks in
'elfin' forest, at an altitude of 1045 m.

37. **Asplenium sp. "H781" ined.**

Epiphyte; rhizome erect, ± 6 mm diameter, the scales mid-brown, narrowly
ovate, to 3 × 0.5 mm, entire, attenuate and ending in a hair-tip, clathrate. Leaves
tufted. Stipe chestnut brown, 17–21 cm long, glabrous except near base. Lamina
thinly coriaceous, narrowly elliptic in outline, 27–28 × 12–13 cm, 1-pinnate, not
proliferous; basal pinnae slightly or not reduced, terminal pinna similar to other
pinnae but more lobed at base; with very few minute scales on lower surface.
Pinnae in 10–12 pairs, narrowly and asymmetrically ovate, to 6.5 × 2.2 cm,
acroscopic base parallel to midrib and with relatively wide lobe, basiscopic base
more cuneate, margin crenulate, attenuate. Rachis brown abaxially, not winged,
with very few subulate scales to 1 mm. Sori in lower pinnae restricted to upper
part of pinna, in more terminal pinnae spread over upper $^2/_3$ of pinna lamina, to
15 per pinna, linear, along the veins, 4–12 mm long; indusium membranous,
entire, 0.4 mm wide.

UGANDA. Kigezi District: Bwindi forest, Ihihizo, Aug. 1998, *Hafashimana* 781!
DISTR. **U** 2; not known elsewhere
HAB. Mature moist evergreen secondary forest; 1520 m
CONSERVATION NOTES. Data deficient (DD)

NOTE. Possibly a new taxon – more material would be interesting. It resembles *A. gemmascens*
but while that taxon is both long-creeping with spaced fronds, and proliferous, this
specimen is neither.

38. **Asplenium mannii** *Hook.*, Sec. Cent. Ferns: t. 60 (1861); Sim, Ferns S. Afr. ed. 2: 174, t. 61 (1915); Tardieu in Mém. Inst. Fr. Afr. Noire, 28: 198, t. 39 fig. 8–9 (1953); Alston, Ferns W.T.A.: 60 (1959); Schelpe, F.Z. Pteridophyta: 187, t. 54e (1970); Burrows, S. Afr. Ferns: 239, map, figs. (1990); Johns, Pterid. trop. East Africa checklist: 65 (1991); Faden in U.K.W.F. ed. 2: 29, t. 172 (1994). Type: Bioko [Fernando Po], *Mann 372* (K!, holo., B!, iso.)

Epiphyte; rhizome short and erect, 1–3 mm diameter, with clathrate dark brown triangular subentire acute rhizome scales to 2.5 mm long. Fronds of two kinds: tufted, erect and not proliferous, thinly coriaceous, 2–16 cm long; and proliferous stolon-like looping green fronds 2–50 cm which are without lamina, start erect and then loop over and root at their tip; pinnae of normal fronds and gemmae of stolon-like fronds alternate. 'Normal' fronds with stipe greyish-green when dried, 1–6.5 cm long, glabrous; lamina dark green or bluish green, narrowly ovate to lanceolate in outline, 1.5–6(–15) × 1–2.5 cm, mostly 1-pinnate but with the lowest pinnae bifid and larger than those above; pinnae up to 6 (very rarely 9) pairs, up to 1.5 cm long, with linear to spatulate lobes, entire, obtuse, glabrous on both surfaces or with sparse thread-like scales to 0.5 mm long beneath; rachis glabrous, narrowly winged. Sori solitary on each spathulate segment, set at a slight angle to the lower part of the vein, oblong, 1.5–2.5 mm long; indusium oblong, membranous, subentire, 0.5–1 mm wide. Fig. 6: 7–8, p. 37.

UGANDA. Toro District: Kibale Forest National Park, near Mpanga R., Sep. 1997, *Lye & Katende* 22931!; Kigezi District: Bwindi/Impenetrable National Park, Aug. 1998, *Hafashimana 508!*; Masaka District: Sese Islands, July 1935, *A.S. Thomas 1376!*
KENYA. Embu District: Irangi Forest Station, Apr. 1972, *Faden et al. 72/187!*; Kericho District: Marinyn tea estate, 4 km SSE of Kericho, June 1972, *Faden et al. 72/328!*; Masai District: Chyulu Hills, main forest, Feb. 1998, *Luke & Luke 5216!*
TANZANIA. Lushoto District: Mafi Hill near headwaters of Kwalukonge stream, Jan. 1985, *Borhidi et al. 85/296!*; Kilosa District: Ukaguru Mts, near Mandege Forest Station, May 1978, *Thulin & Mhoro 2734!*; Njombe District: Livingstone Mts, Madunda Mission, Feb. 1961, *Richards 14209!*
DISTR. **U** 2, 4; **K** 4–7; **T** 2–4, 6–8; widespread in tropical Africa, from Sierra Leone to Ethiopia to Zimbabwe; Madagascar
HAB. Low, medium and high epiphyte in moist forest; may be locally common and cover entire tree-trunks; (?840–)1100–2500 m
CONSERVATION NOTES. Widespread; least concern (LC)

NOTE. One of the most easily recognizable species in *Asplenium.*

39. **Asplenium blastophorum** *Hieron.* in E.J. 46: 378 (1911); Alston, Ferns W.T.A.: 59 (1959); Schelpe, F.Z. Pteridophyta: 183 (1970); Burrows, S. Afr. Fcrns: 251, map, figs. (1990); Johns, Pterid. trop. East Africa checklist: 62 (1991); Faden in U.K.W.F. ed. 2: 29 (1994). Type: Sudan, Niamniam/Monbuttu border, *Schweinfurth 3295* (B!, lecto., BM!, K (not found), iso.), chosen by ?Burrows

Terrestrial or lithophyte; rhizome short-creeping, to 7 mm diameter, with dark brown subulate entire rhizome scales up to 5 mm long. Fronds shortly spaced, arching, narrowly deltoid, proliferous at base of terminal pinna (and sometimes on pinna costa) near the apex, thinly coriaceous. Stipe matt black, 12–35 cm long, set with brown subulate scales similar to rhizome scales, glabrescent. Lamina ovate to deltate, 15–48 × 8–20 cm, 2-pinnate to 3-pinnatifid near the base of the lamina, the lowest pinnae longest, apical pinna similar to upper pinnae. Pinnae dark green above, ovate or lanceolate in outline, up to 11 × 4.5 cm, unequally cuneate at the base, attenuate, progressively deeply divided towards the base into obcuneate or rhombic pinnules with the outer margins sharply serrate and irregularly shallowly incised, free pinnules with cuneate to narrowly cuneate bases, glabrous except for small lanceolate dark scales near the extreme base.

Rachis matt brown, with occasional dark subulate scales, sulcate ventrally. Sori along the veins, linear, 4–30 mm long; indusium linear, narrow, membranous, entire, 0.3–0.4 mm wide. Fig. 7: 1–2, p. 42.

UGANDA. Toro District: Mwamba Forest, Dec. 1957, *Allen* 3687!
KENYA. North Kavirondo District: Malava Forest, Nov. 1969, *Faden & Evans* 69/2040!
TANZANIA. Morogoro District: NE Uluguru Mts, Kinole, Apr. 1970, *Pócs* 6165/s! & idem, Sep. 1970, *Faden et al.* 70/672!; Iringa/Morogoro District: Mwanihana Forest reserve above Sanje, Sep. 1984, *D.W. Thomas* 3688!
DISTR. U 2; K 5; T 6, 7; tropical Africa from Ivory Coast to Sudan and south to South Africa
HAB. Moist forest; 900–1600 m
CONSERVATION NOTES. Widespread; least concern (LC)

NOTE. I agree with Faden that in habit and cutting of the fronds this species most closely resembles the non-proliferous *A. buettneri*, which however has more dissected pinnae. However, in Zambia there are populations without gemmae – fide Kornaś.

40. **Asplenium udzungwense** *Beentje* **sp. nov.** a *A. bugoiense* differt laminis minus dissectis, lobo basali fere libero. Typus: Tanzania, Iringa District: Udzungwa Mts, above Sanje waterfalls, Nov. 1995, *de Boer, Bosch, Johns & Schippers* 754 (K!, holo.)

Rhizome erect, short and thin, the scales pale brown, ovate, to 3.5 × 1.5 mm, entire, attenuate, clathrate. Leaves tufted. Stipe 5–18 cm long, with few to many scales similar to those of rhizome but decreasing in size. Lamina herbaceous, narrowly elliptic in outline, 12–29 × 3–10 cm, 2-pinnatifid or 3-pinnatifid, proliferous just below terminal pinna; basal pinnae slightly or not reduced, terminal pinna gradually decrescent; glabrous or with few minute scales on lower surface. Pinnae in 13–21 pairs, ovate, to 5 × 1.2 cm, obtuse, with a singe large acroscopic lobe to 12 × 7 mm and slightly 2–3-lobed at apex, main pinna crenate to almost lobed with teeth to 2 mm (lobed to about halfway in *Luke* 8063). Rachis green, narrowly winged, with very few dark brown clathrate lobed scales to 0.7 mm. Sori 8–20 per pinna, linear, along the veins, 1–5 mm long; indusium membranous, entire, 0.3 mm wide. Fig. 7: 3–4, p. 42.

TANZANIA. Morogoro District: Uluguru Mts, Mwere valley, Sep. 1970, *Faden et al.* 70/582! & 70/617!; Iringa District: Udzungwa Mountains, above Sanje waterfalls, Nov. 1995, *de Boer et al.* 754! & Point 236, Oct. 2001, *Luke et al.* 8071! & 8063!
DISTR. T 6, 7; not known elsewhere
HAB. Montane forest; 1000–1550 m
CONSERVATION NOTES. Though the habitat is threatened in the Uluguru Mts, it is probably safe in the Udzungwa Mts; Vulnerable (VU-D2) due to the Uluguru threat coupled to the few locations.

NOTE. This slightly resembles *A. bugoiense*, differing in the less-dissected lamina, though with a well-developed acroscopic lobe that is almost free; or a proliferous *A. auritum*, and again the well-developed acroscopic pinna lobe is characteristic.

41. **Asplenium protensum** *Schrad.* in Gött. Gel. Anz. 1818: 916 (1818); Sim, Ferns S. Afr. ed. 2: 149, t. 51 (1915); Tardieu in Mém. Inst. Fr. Afr. Noire, 28: 183, t. 37 fig. 5–6 (1953); Alston, Ferns W.T.A.: 57 (1959); Tardieu, Fl. Camér. 3, Ptérid.: 194, t. 29/5–6 (1964); Schelpe, F.Z. Pteridophyta: 179 (1970); Burrows, S. Afr. Ferns: 232, map, figs. (1990); Johns, Pterid. trop. East Africa checklist: 66 (1991); Faden in U.K.W.F. ed. 2: 28, t. 172 (1994). Type: South Africa, Cape Province, *Hesse* s.n. (?LE, holo.) (not at B)

Low-level epiphyte, on wet rocks or terrestrial; rhizome erect or sometimes shortly creeping, smelling of wintergreen (methyl salicylate) when cut, up to 5 mm diameter, with dark brown ovate-triangular clathrate minutely pseudoserrate rhizome scales 1–1.5(–2) mm long, with pale lacerate margins. Fronds tufted, arching, narrowly

FIG. 7. *ASPLENIUM BLASTOPHORUM* — **1**, pinnae × ¹/₂; **2**, subterminal gemma × ²/₃. *ASPLENIUM UDZUNGWENSE* — **3**, habit, frond × ²/₃; **4**, pinnae × ²/₃. *ASPLENIUM PROTENSUM* — **5**, habit, frond × ¹/₃; **6**, pinna × ²/₃; **7**, sori × 1¹/₂. *ASPLENIUM DREGEANUM* — **8**, habit × ²/₃. *ASPLENIUM BUGOIENSE* — **9**, pinnae × ²/₃; **10**, detail part pinnae × 2; **11**, subterminal gemma × ²/₃. 1, 2, from *de Boer* 730; 3 from *de Boer* 754; 4, from *Luke* 8071; 5, 7, from *Verdcourt & Moggi* 2503; 6, from *Tweedie* 1824; 8, from *de Boer* 752; 9–11, from *Faden et al.* 72/289. Drawn by Juliet Williamson.

oblong-elliptic, herbaceous, proliferous near the apex of the frond (rarely not proliferous). Stipe shiny dark brown, 4–20 cm long, densely set with ovate to broadly ovate clathrate pale scales of various sizes up to 1.5 mm long (–2.5 mm on crozier stipes) together with multicellular hairs. Lamina very narrowly elliptic in outline, 15–95 × 3–12 cm, 2-pinnatifid, lower pinnae (much) reduced, apical pinna linear and lobed. Pinnae in 24–56 pairs, lanceolate, rhombic, deeply but narrowly lobed to $^1/_3$–$^2/_3$ of the way, 2–7 × 0.8–1.7 cm, base unequally cuneate, acroscopic basal lobe the largest and often parallel to the rachis, pinnately divided into linear-oblong acute 2-fid or 3-fid or broadly cuneate and deeply incised lobes, lobes oblique, truncate and bifid, thinly pubescent above and below especially near the base, densely so along the costa, dorsally with clathrate ovate brown scales up to 1 mm long; veins flabellate, costa may be difficult to see or pronounced. Rachis green above, dark brown to black at sides and beneath, pubescent with whitish hairs and with occasional ovate scales. Sori 10–22, in 2 rows close to costa and set at 15–25°, linear and very narrowly oblong, 2–6.5 mm long; indusium linear, membranous, entire, 0.3–0.5 mm wide. Fig. 7: 5–7, p. 42.

UGANDA. Toro District: Kibale National Park, SW of Ngogo camp, June 1997, *Poulsen & Nkuutu* 1304!; Ankole District: Kalinzu Forest, 4 km NW of saw mill, Sep. 1969, *Faden et al.* 69/1158!; Elgon, Dec. 1950, *van Someren* 581!

KENYA. Turkana District: Murua Nysigar [Ngithigerr], Sep. 1963, *Paulo* 1013!; North Kavirondo District: Kakamega Forest near Forest Station, Oct. 1981, *Gilbert & Mesfin* 6675!; Machakos District: Chyulu Hills, main forest, camp 3, Oct. 1997, *Luke & Luke* 4813!

TANZANIA. Ngara District: Rusumo Falls, Mar. 1960, *Tanner* 4792!; Kilimanjaro, above Kilimanjaro Timbers, Sep. 1993, *Grimshaw* 93/718!; Mbeya District: Sawago forest, 2 km W of Igoma, June 1992, *Gereau et al.* 4589!

DISTR. **U** 2, 3; **K** 2–6; **T** 1–3, 7; much of tropical Africa, south to Zambia, Malawi, Mozambique, Zimbabwe, South Africa; Mascarene Is.

HAB. Moist forest, usually near streams or on rocks in stream; 1250–2900(–3300) m

CONSERVATION NOTES. Widespread; least concern (LC)

NOTE. This taxon resembles *A. pellucidum* closely but differs in being proliferous, as well as in having much smaller rhizome scales.

42. **Asplenium dregeanum** *Kunze* in Linnaea, 10: 517 (1836); Alston, Ferns W.T.A.: 59 (1959); Schelpe, F.Z. Pteridophyta: 184 (1970); Pic. Serm. in B.J.B.B. 55: 130 (1985); Burrows, S. Afr. Ferns: 236, map, figs. (1990); Johns, Pterid. trop. East Africa checklist: 63 (1991); Faden in U.K.W.F. ed. 2: 29 (1994). Type: South Africa, Transkei, between Umzimvubu [Omsamwubo] and Umsikaba [Omsamcaba] Rs., *Drège* 158 (LZ†, holo.; BM, lecto. – not found, HBG, iso., photo.!)

Epiphyte, lithophyte or terrestrial; rhizome erect, to 5 mm diameter, with brown lanceolate to narrowly ovate dark brown clathrate rhizome scales to 3 (?–5) mm long with pale entire margin. Frond tufted, erect to arching, herbaceous, proliferous below the apex. Stipe greyish-brown, 4–17 cm long, with narrow green wings when fresh, usually less than half the length of the lamina, with occasional clathrate scales similar to those of the rhizome becoming more frequent towards the base. Lamina very narrowly oblong to narrowly oblong-elliptic in outline, 10–39 × 2.5–6 cm, 2-pinnatisect to 3-pinnatifid on basal acroscopic lobes, lowest pinnae hardly reduced, apical segment deeply pinnatifid with 3–5 lobes. Pinnae 13–32 pairs, petiolate, rhombic-oblong or narrowly oblong in outline, up to 3.5 × 1.1 cm, variously acute-auriculate with the acroscopic basal lobe 2–4-fid, base dimidiate with basiscopic margin formed by midrib for $^1/_3$ of pinna, most of the other segments very narrowly oblong-obtuse (some larger segments 2-fid) up to 6 × 1.5 mm, subglabrous with a few minute substellate scales on the veins. Rachis matt-greyish-green when dry, with occasional pale brown minute substellate scales. Sori one central per pinna lobe, broadly elliptic at maturity, 2–3 mm long; indusium elliptic, membranous, entire, to 1 mm wide. Fig. 7: 8, p. 42.

UGANDA. Ankole District: Kasyoke-Kitomi forest, Nzozi, June 1998, *Hafashimana* 642!; Kigezi District: Bwindi Impenetrable National Park, Aug. 1998, *Hafashimana* 699!; Mengo District: 1.5 km NE of Nansagazi, Sep. 1969, *Lye et al.* 3887!

KENYA. Meru District: Nyambeni [Jombeni] Hills, Aug. 1949, *H.D. van Someren* 585!; North Kavirondo District: along Isiukhu R. on Kambiri–Vihiga road, Dec. 1969, *Faden & Rathbun* 69/2103!; Teita District: Kasigau, Apr. 1969, *Faden et al.* 69/468! & idem, June 1998, *Luke & Luke* 5352!

TANZANIA. Bukoba District: Minziro forest, Sep. 1952, *Procter* 112!; Morogoro District: North Uluguru Forest Reserve, Palu, Dec. 1993, *Kisena* 991!; Rungwe District: Livingstone Mts, SE of Bumbigi, Mar. 1991, *Gereau & Kayombo* 4170!

DISTR. **U** 1, 2, 4; **K** 4, 5, 7; **T** 1, 3, 4, 6–8; tropical Africa from Sierra Leone to East Africa and south to South Africa; Madagascar, Comoro Is.

HAB. Moist forest, either a low epiphyte, terrestrial, or on mossy rocks, often by stream-sides; 850–1900(–2100) m

CONSERVATION NOTES. Widespread; least concern (LC). Four collections from Kenya, but not uncommon in Uganda and Tanzania.

SYN. *A. brachypterum* Houlston & Moore in Gard. Mag. Bot. 3: 260 (1851). Type: Sierra Leone, no details (often cited as *brachypteron*)
　　A. gracile Peter, F.D.O.-A.: 73 (1929) & Descr.: 5, t. 5.2 (1929); Johns, Pterid. trop. East Africa checklist: 63 (1991), *non* D.Don (1825). Type: Tanzania, without locality, *Peter* 41361 (B, holo., not found)
　　A. dregeanum Kunze subsp. *brachypterum* (Houlst. & T.Moore) Pic.Serm. in B.J.B.B. 55: 130 (1985)

NOTE. While the type of *A. gracile* does not seem to be present in the Berlin herbarium, another specimen from the Uluguru Mts, Mlari stream, *Peter* 32241! has been seen; this was cited by Peter with the *nomen nudum*.
　　De Boer et al. 784 from **T** 7, Sanje Falls, has the pinnae longer and more regular than usual.

43. **Asplenium preussii** *Hieron.* in Z.A.E. 2: 9 (1910); Alston, Ferns W.T.A.: 59 (1959); Burrows, S. Afr. Ferns: 234, map, figs. (1990); Johns, Pterid. trop. East Africa checklist: 66 (1991). Type: Cameroon, Mt Cameroon, Buea, *Preuss* 584 (B!, lecto.; B!, iso.), chosen by Tardieu (1964)

Terrestrial, epiphyte or lithophyte; rhizome erect, 5–10 mm diameter, with dark brown subulate subentire concolorous rhizome scales 3–7 × 2 mm, hair-tipped. Frond tufted, arching, thinly coriaceous, to 60 cm long, proliferous at the base of the deeply pinnatifid lanceolate apical segment. Stipe green when fresh, greyish-green when dried, 8–20 cm long, with sparse dark brown lanceolate clathrate scales up to 2 mm long. Lamina 22–43 × 4.5–13 cm, 2-pinnatisect, oblong lanceolate in outline, acuminate, basal pinnae hardly reduced; pinnae variable, from shallowly incised to pinnatifid almost to costa, up to 8.5 × 2 cm, petiolate, lanceolate-attenuate, unequally cuneate at the base, deeply pinnatifid into linear or very narrowly oblong-acute or oblanceolate 2-fid lobes up to 8 mm long, basal acroscopic lobe broadly cuneate, shallowly incised in the upper $\frac{1}{2}$ into 4–7 acute short lobes, glabrous except for a few substellate dark brown minute scales on the lower surface. Rachis matt-greyish-green with minute substellate to very narrowly lanceolate dark brown scales. Sori set along the veins, one per lobe, linear, 2–7 mm long, slightly curved; indusium very narrowly oblong, membranous, entire.

UGANDA. Ruwenzori, Namwamba Valley, 5 Jan. 1935, *G. Taylor* 2866!

KENYA. Kakamega District: Kakamega Forest, 5 km SE of Forest Station, Nov. 1969, *Faden et al.* 69/2018!

TANZANIA. Morogoro District: S Uluguru Mts, slopes above Simbini, Mar. 1971, *Pócs et al.* 6418R!

DISTR. **U** 2; **K** 5; **T** 6; sporadic throughout tropical Africa, to South Africa

HAB. Moist forest; 1400–2350 m

CONSERVATION NOTES. Widespread; least concern (LC)

SYN. *A. pseudoauriculatum* Schelpe in Bol. Soc. Brot. ser. 2, 41: 206 (1967) & F.Z. Pteridophyta: 185 (1970); Johns, Pterid. trop. East Africa checklist: 66 (1991); Faden in U.K.W.F. ed. 2: 29 (1994). Type: Mozambique, Manica e Sofala, Garuso, Jaegersberg, *Schelpe* 5626 (BOL, holo.; BM!, iso.)

NOTE. A variable species. I agree with Burrows that the finely divided extreme form resembles *A. dregeanum*. *A. bugoiense* is also close, differing mainly in the dissection details; there are intermediates such as *Faden* 69/2117. When specimens lack gemmae, they will key to the *A. rutifolium/loxoscaphoides* group.

44. **Asplenium bugoiense** *Hieron.*, Z.A.E.: 10, t. 2a, b (1910); Johns, Pterid. trop. East Africa checklist: 62 (1991); Faden in U.K.W.F. ed. 2: 29, t. 173 (1994). Type: Rwanda, "Bugoier forest SE of Kissenje", *Mildbraed* 1460 (B!, holo.)

Terrestrial, occasionally a low epiphyte or lithophyte; rhizome erect or short-ascending, to 9 mm diameter, with dark brown rhizome scales with cordate base, ovate-oblong or elongate-deltoid, acute to long-acuminate, margin pale brown, entire or rarely fimbriate, dark, pellucid, to 6 × 2.5 mm. Fronds tufted or occasionally shortly spaced, often with subterminal gemmae; stipe dark green above, black beneath, 11–23 cm long, with scales similar to rhizome scales (but smaller) dense near base and sparse elsewhere. Lamina dark green, ovate-lanceolate in outline, (20–)30–50 × 8–20 cm, 3-pinnatisect (to slightly 4-pinnatisect), acuminate, the apical part gradually decrescent; pinnae shortly stalked, in 13–23 pairs, the basal ones hardly smaller and (sub)opposite, the upper ones alternate, elongate-deltoid, to 8 × 4(–10 × 5, fide protologue) cm, 2-pinnatisect, acroscopic base truncate, basiscopic base cuneate; pinnules in 6–12 pairs; the ultimate segments falcate-oblong and to 8 × 1.5 mm; scales sparse, as on rachis. Rachis narrowly winged, glabrous or with sparse dark brown scales to 2 mm, the margins with hair-lobes. Sori 1(–2) per segment, on the vein, 1.5–4 mm long; indusium membranous, entire, 0.4–0.5 mm wide. Fig. 7: 9–11, p. 42.

UGANDA. Karamoja District: Mt Kadam [Debasien], no date, *Eggeling* 2690!; Kigezi District: Bwindi-Impenetrable forest, Nov. 1989, *Rwaburindore* 2909!; Mt Elgon, Oct. 1961, *Rose* 2005! & Dec. 1966, *Wesche* 566!
KENYA. Kiambu District: Kikuyu Escarpment forest, Ndiara waterfall, Aug. 1974, *Faden et al.* 74/1346!; Mt Kenya, above Castle Forest Station, Oct. 1979, *Gilbert* 5801!; Kericho District: Kimugung R. 5 km NW of Kericho, June 1972, *Faden et al.* 72/289!
TANZANIA. Kilimanjaro, Old Moshi, above Kidia, Dec. 1995, *Hemp* 911! & above Uru-east, Nov. 1996, *Hemp* 1109!
DISTR. **U** 1–3; **K** 3–5; **T** 2; Rwanda
HAB. Moist forest and bamboo forest, nearly always in damp sites such as stream-banks, often associated with *Cyathea*; may be locally common; 1650–2550(–3000?) m
CONSERVATION NOTES. Widespread, Least Concern (LC)

SYN. *Asplenium sp. C* of Faden in U.K.W.F. ed. 2: 29 (1994); Johns, Pterid. trop. East Africa checklist: 68 (1991)
Asplenium sp. D of Faden in U.K.W.F. ed. 2: 29 (1994); Johns, Pterid. trop. East Africa checklist: 68 (1991)

NOTE. The much-lobed first acrocopic pinnule also features in several specimens of *A. loxoscaphoides*.
The specimens from **U** 2 differ slightly in having the first acroscopic pinnule less developed. A specimen from Kenya, Kiambu District: Aberdares, Royi, Mar. 1921, *Gardner* 1266! differs in the slightly narrower ultimate segments, and the distal, unlobed pinnules of the middle pinnae often at an angle of 70–90° to the costa, while those of *A. bugoiense* are usually at 45–60°. This is *Asplenium sp. C* of Faden. *Gillett* 16703 from **K** 5 Kaimosi looks like this taxon but lacks frond tips, so identification is not quite sure. Specimen (*Faden* 69/2019!) sterile, but most likely *A. bugoiense*; frond cutting differs slightly, but not much.

45. **Asplenium linckii** *Kuhn*, Fil. Deck.: 22 (1867) & Fil. Afr.: 22 (1868); Schelpe, F.Z. Pteridophyta: 183 (1970); Burrows, S. Afr. Ferns: 251, map, figs. (1990); Johns, Pterid. trop. East Africa checklist: 65 (1991); Faden in U.K.W.F. ed. 2: 29 (1994). Type: Tanzania, Kilimanjaro, [Dschagga], *Kersten* in *von der Decken* 40 (B!, holo.)

Terrestrial, low epiphyte, or lithophyte; rhizome shortly creeping to suberect, up to 5 mm diameter, with shining dark brown clathrate narrowly triangular entire rhizome scales up to 6 × 0.3–1 mm, attenuate and ending in a hair-tip. Frond shortly spaced or tufted, arching, firmly herbaceous, generally 30–50 cm long, not proliferous. Stipe matt brown to black, 9–40 cm long, with appressed scales similar to those of the rhizome near the base, subglabrous above. Lamina dark green on upper surface, ovate-triangular in outline, (8–)20–33(–60) × (4–)10–27 cm, 3–4-pinnate to 5-pinnatifid, finely dissected, basal pinnae the largest, apex decrescent. Pinnae in 15–20 pairs, broadly ovate or triangular in outline, up to 17 × 10 cm, ultimate segments wedge-shaped or narrowly obcuneate, up to 8–10 × 2–4 mm, apex doubly serrate-truncate with teeth to 2 mm long, glabrous on both surfaces or with a few minute scales beneath. Rachis and secondary rachises dark brown, with sparse pale brown ovate-acuminate clathrate scales 1–2 mm long. Sori 1–3 per ultimate segment, 1.5–6 mm long; indusium linear, membranous, entire, 0.3–0.4 mm wide. Fig. 8: 1–3, p. 47.

UGANDA. Toro District: Ruwenzori, Mobuku Valley, *Loveridge* 274!; Kigezi District: Bwindi
   Impenetrable Forest, Sep. 1961, *Rose* 10315! & Oct. 1961, *Lind* 3191!
KENYA. Kiambu District: Kieni Forest 8 km E of Kieni, June 1986, *Beentje & Mungai* 2928!; Embu
   District: Castle Forest Station, Dec. 1972, *Gillett & Holttum* 20095!; Teita District: Taita Hills,
   Ngangao, May 1985, *NMK Taita Hills Expedition* 270!
TANZANIA. Kilimanjaro: above Kilimanjaro Timbers, June 1993, *Grimshaw* 93/149!; Morogoro
   District: Nguru Mts, above Maskati, Mar. 1988, *Bidgood et al.* 510!; Iringa District: Udzungwa
   Mountain National Park, Mt Luhombero, Sep. 2000, *Luke et al.* 6821!
DISTR. **U** 2; **K** 3, 4, 7; **T** 2, 3, 6, 7; Congo-Kinshasa, Rwanda, Burundi, Malawi, Zimbabwe
HAB. Deeply shaded moist forest; (1600–)1700–2300(–2700) m
CONSERVATION NOTES. Widespread; least concern (LC)

SYN. *A. daubenbergeri* Rosenst. in F.R. 4: 2 (1907). Type: Tanzania, Kilimanjaro, Kibosho,
   Daubenberger 43 (B!, holo., BM!, K!, iso.)
   A. albersii Hieron. var. *eickii* Hieron. in E.J. 46: 381 (1912). Type: Tanzania, Usambara, Kwai,
   Eick 119 (B!, holo.)
   A. decompositum Peter in F.D.O. A.: 79 (1929) & Descr: 7, t. 2.5–2.6 (1929); Johns, Pterid.
   trop. East Africa checklist: 62, 63 (1991). Type: Tanzania, Kilimanjaro, above Moshi,
   Peter 1305, 1341 (both B!, syn.), **syn. nov**. Note: in B someone has written 'lectotype' on
   Peter 1305, but without leaving their name
   A. albersii Hieron. var. *kirkii* of Johns, Pterid. trop. East Africa checklist: 61 (1991) –
   misreading of *eickii*

NOTE. The name has been spelled as *linkii* in some of the literature, but the protologue uses *linckii* with a 'c'. Linck was a collector on the von der Decken expedition.
   There is a link with *A. albersii*, a 3-pinnatisect taxon. *A. albersii* var. *eickii* has very narrow rhizome scales, about 0.3 mm wide, but without the paler margin such as *albersii* is supposed to have; the lamina is 4-pinnatisect, and the sori 1.5–3 mm long. The protologue already stated "this variety resembles in habitat more *A. linkii* [sic] Kuhn, as the lamina shape is ovate; but rhizome scales are typically *A. albersii*." With this last statement I disagree. Several specimens from **U** 2 that are 3-pinnatifid or 3-pinnatisect to 3-pinnate could be intermediate, but as they lack basal parts it is impossible to be quite certain: *Thomas* 2039 from Butale, *Esterhuysen* 25313 from Bujuku Valley, *Thomas* 1431 from Bwamba Pass, *Thomas* 2424 from Mt Mgahinga, *Rwaburindore* 2894. These specimens are from slightly higher altitudes, 2300–2600 m.
   Burrows states this resembles a finely-divided form of *A. aethiopium* or *A. uhligii*. I agree; *Holst* 3865 only has the lowermost 2nd-order pinnules divided, and so is only just 4-pinnatisect. Chaerle, in his unpublished Ph.D. thesis, keeps *A. decompositum* separate, distinct in its lamina dissection and thinner rhizome, as well as lamina shape.

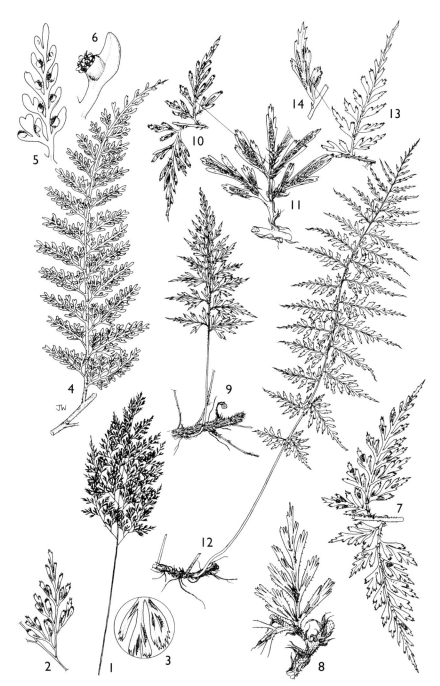

FIG. 8. *ASPLENIUM LINCKII* — **1**, frond × ¹/₆; **2**, 2nd branching pinna × ²/₃; **3**, sori × 1¹/₂. *ASPLENIUM HYPOMELAS* — **4**, pinna × ²/₃; **5**, part pinna × 2; **6**, sori × 6. *ASPLENIUM ACTINIOPTEROIDES* — **7**, pinnae × ²/₃; **8**, sori × 1¹/₂. *ASPLENIUM UHLIGII* — **9**, habit × ¹/₃; **10**, pinnae × ²/₃; **11**, detail pinnae × 1¹/₂. *ASPLENIUM PRAEGRACILE* — **12**, habit, × /₃; **13**, pinna upper surface × ²/₃; **14**, sori × 1¹/₂. 1, 2, from *Loveridge* 274; 3, from *F. Rose* 10317; 4–6, from *Drummond & Hemsley* 2805; 7, from *Vesey Fitzgerald* 5699; 8, from *Pichi Sermolli* 5143; 9–11, from *Kenya Exploration Soc.* 154; 12, 13, from *Noble* 13; 14, from *Wood* 224. Drawn by Juliet Williamson.

46. **Asplenium sp. "S618" ined.**

Rhizome scales pale brown, ovate, to 5 × 1.6 mm, entire or with few minute side-lobes, attenuate, clathrate. Leaves tufted. Stipe 7–14 cm long, with few to many scales similar to those of rhizome but decreasing in size. Lamina herbaceous, narrowly elliptic in outline, 30–60 × 10–18 cm, 3-pinnate to 4-pinnatifid on acroscopic pinna lobe, not proliferous, with few minute scales on lower surface. Pinnae in ± 18–19 pairs, narrowly ovate, to 10 × 3 cm, attenuate, with up to 13 pinnules on each side, 2–8-lobed with narrow spatulate lobes, most with notched apex. Rachis black abaxially, unwinged, with scattered clathrate dark brown scales with sparse filiform side-lobes. Sori 1 per segment, linear, almost marginal and opening outwards, 2.5–4 mm long; indusium membranous, entire, 0.5 mm wide.

KENYA. Embu District: Embu Forest Station, Aug. 1949, *H.D. van Someren* 618! & Kwale District: Shimba Hills, Mar. 1941, *H.D. van Someren* 123!
DISTR. **K** 4, 7; not known elsewhere
HAB. Wet forest, on rocks and fallen trunks by stream; altitude unknown
CONSERVATION NOTES. Data deficient (DD), but not collected for a long time

NOTE. Keys to *A. loxoscaphoides* but completely different in its dissection – *loxoscaphoides* only has the acrocopic basal pinnules lobed to 3rd order; S618 has most pinnules lobed! As both sheets are incomplete, I am reluctant to describe a new species, but I am fairly certain that this is new.

47. **Asplenium hypomelas** *Kuhn*, Fil. Afr.: 104 (1868); Alston, Ferns W.T.A.: 59 (1959); Schelpe, F.Z. Pteridophyta: 187, t. 54a (1970); Burrows, S. Afr. Ferns: 239, map, figs. (1990); Johns, Pterid. trop. East Africa checklist: 63 (1991); Faden in U.K.W.F. ed. 2: 29 (1994). Type: Bioko [Fernando Po], *Mann* 448 (K!, holo.; B!, iso.)

Epiphyte, occasionally terrestrial, rarely lithophytic; rhizome erect, to 20 cm long and 15 mm diameter, with narrowly lanceolate attenuate glossy brown sparsely fimbriate rhizome scales up to 13 × 2.5 mm long with short hair-points. Frond tufted, arching, not proliferous, dark-green when fresh, herbaceous. Stipe grey-green to black, 18–45 cm long, sulcate, covered with hair-tipped scales at first, becoming glabrous with age. Lamina glossy dark to bright green, lanceolate to broadly lanceolate, 50–100 × 19–46 cm, 4-pinnatisect, the acroscopic pinnules more developed, the basal pinnae somewhat reduced. Pinnae up to 25 pairs, lanceolate in outline, up to 26 × 9 cm, the ultimate segments narrowly obovoid to spatulate, 0.8–1.5 mm wide, acute, unequally expanded around the sori, glabrous on both surfaces or with sparse broad-based but rapidly attenuate scales to 2 mm long. Rachis and secondary rachises black above, green beneath, with brown broad-based but rapidly attenuate and long hair-pointed scales; secondary rachises narrowly winged for most of their length. Sori on the acroscopic margins of the ultimate lobes, cup-shaped and protruding from the margin, broadly oblong, 1–1.5 mm long; indusium broadly elliptic to broadly oblong, membranous, entire, to 1 mm long/wide. Fig. 8: 4–6, p. 47.

UGANDA. Toro District: Bwamba Pass, Nov. 1935, *A.S. Thomas* 1459!; Kigezi District: Bwindi Impenetrable forest, Ishasha gorge, Apr. 1998, *Hafashimana* 513!; Masaka District: Sesse Is., 1904, *Dawe* 64!
KENYA. Mt Kenya, Castle Forest Station, Dec. 1966, *Kabuye* 48!; Kericho District: SW Mau forest along Kiptiget R., 16 km SSE of Kericho, June 1972, *Faden et al.* 72/346!; Teita District: Sagala Hill, June 1985, *Taita Hills Expedition* 1083!
TANZANIA. Lushoto District: Shume-Magamba Forest Reserve, Mar. 1987, *Kisena* 628!; Kilosa District: Ukaguru Mts, below Mnyera peak, July 1972, *Pócs & Mabberley* 6742/C!; Iringa District: Udzungwa Mountain National Park, Oct. 2002, *Luke et al.* 9224!
DISTR. **U** 2–4; **K** 3–5, 7; **T** 2–4, 6–8; West tropical Africa to Ethiopia, and S to Malawi, Mozambique and Zimbabwe
HAB. Moist forest, often along rivers, often epiphytic on tree ferns, less often on other trees, sometimes terrestrial, rarely on rock; 800–2400 m

CONSERVATION NOTES. Widespread; least concern (LC)

SYN. *Davallia nigrescens* Hook., Sec. Cent. Ferns: t. 93 (1861), *non Asplenium nigrescens* Bl. (1828). Type as for *A. hypomelas*
  Loxoscaphe nigrescens (Hook.) Moore, Ind. Fil.: 297 (1861); Pic.Serm. in B.J.B.B. 55: 123 (1985)
  Davallia hollandii Sim in Trans. S. Afr. Phil. Soc 16: 274, t. 4 (1906). Type: Mozambique, Penhalonga, *Holland* s.n. (NBG, isotype; PRE, holotype).
  Asplenium floccigerum Rosenst. in F.R. 4: 3 (1907). Type: Tanzania, Kilimanjaro, 3000-4000 m, *Daubenberger* s.n. (B, holo., not found)
  A. hollandii (Sim) C.Chr. Ind. Fil., Suppl.: 11 (1913)
  A. spathulatum Peter, F.D.O.-A. Descr: 8 (1938), *non* Baker 1894. Type: Tanzania, Morogoro District: Uluguru Mts, SE of Schlesien Mission, *Peter* 6858 (B!, holo.)
  Loxoscaphe spathulata Pic.Serm. in Webbia 32: 77 (1977)
  Asplenium morogorense Viane in Biol. Jaarb. Dodonaea 59: 161 (1991), *nom. nov.* Type as for *A. spathulatum*; **syn. nov.**

NOTE. Not easily confused with any other *Asplenium*.
  I have made *A. spathulatum* of Peter = *A. morogorense* a synonym; the differences indicated by Peter do not hold true, as they are based on the incomplete type – consisting only of the upper part of the frond, and so the shorter stipe (which is not present) and the fewer pinnae (based on an incomplete frond) cannot be upheld.

48. **Asplenium actiniopteroides** *Peter* in F.D.O.-A.: 79 (1929) & Descr.: 7, t. 3/3–4 (1929); A.V.P.: 28 (1957); Johns, Pterid. trop. East Africa checklist: 61 (1991); Faden in U.K.W.F. ed. 2: 30 (1994). Type: Tanzania, Kilimanjaro, Peters' Hut to Bismarck Hut, *Peter* 1084 (B!, holo.; K!, iso.)

Lithophyte, terrestrial or occasionally epiphyte; rhizome long-creeping, 2–5 mm diameter, with dense dark brown lanceolate subentire scales 3–6 × 0.8–1.2(–2.8) mm, ending in a hair-tip. Fronds spaced, sparse, erect, not proliferous. Stipe 10–30 cm long, with a few to many scales 1.5–2 mm long, usually also with rather dense stalked glands. Lamina dark or dull green, sub-coriaceous, narrowly ovate in outline, (6–)7–28 × 3.5–18 cm, 2-pinnate to 3-pinnatifid, rarely 4-pinnatifid; pinnae in 5–19 pairs, obliquely ovate or trapezoid, to 5 × 2.5 (–9 × 4.5) cm, the basal ones largest, apex decrescent; pinnules deltate, pinnatisect, with 2–7 segments, the ultimate segments with 2–4 narrowly linear obtuse apices to 0.4 mm wide, with occasional scales; rachis rather densely scaly with long-attenuate scales, often (?always) with rather dense stalked glands – but these often absent on young/infertile fronds. Sori 1–2(–3) per pinnule or lobe, linear, 1.5–3.5 mm; indusium present, entire, 0.2–0.5 mm wide. Fig. 8: 7–8, p. 47.

UGANDA. Ruwenzori Mts, Oct. 1905, *Dawe* 568!; Mt Elgon, above Butandiga ridge, Mar. 1951, *G. Wood* 173! & Sasa trail, Mar. 1997, *Wesche* 1071!
KENYA. Mt Elgon, Feb. 1931, *Lugard* 688! & crater, May 1948, *Hedberg* 954!; Nyandarua/Aberdare Mt., Feb. 1959, *H.D. van Someren* 1054!
TANZANIA. Kilimanjaro, Shira plateau below Simba camp, Dec. 1993, *Grimshaw* 93/1262!; Arusha District: Mt Meru: W slopes above Olkakola, Oct. 1948, *Hedberg* 2303! & Meru crater, Mar. 1970, *Vesey-Fitzgerald* 6588!
DISTR. **U** 2, 3; **K** 3; **T** 2; Rwanda; probably also occurs in Congo-Kinshasa; restricted to Ruwenzori, Virunga, Mt Elgon, Nyandarua/Aberdares, Mt Meru and Kilimanjaro
HAB. From the upper part of the forest through Hagenia woodland and heath zone to upper moorland, where restricted to rock crevices; terrestrial, epiphytic or lithophytic; 2500–4250 m
CONSERVATION NOTES. Fairly widespread with a wide altitude range; Least Concern (LC)

SYN. *A. goetzei* Hieron. var. *major* Hieron in Z.A.E. 2: 19, 21 (1910). Type: Rwanda, Karisimbi, *Mildbraed* 1590 (B!, lecto., chosen by Pichi Sermolli)
  A. majus (Hieron.) Pic.Serm. in Webbia 37(1): 134 (1983); Pic. Serm. in B.J.B.B. 55: 143 (1985); Schippers in Fern Gaz. 14, 6: 202 (1993), **syn. nov.**

NOTE. This taxon resembles *A. uhligii* but the lamina is usually ovate-deltoid and more finely dissected; dark scales and glandular rachis are best distinguishing features.

Gland cover is variable: they are missing in *Hedberg* 1289, which otherwise is a perfect match; there are few on *Hedberg* 2303 and *Grimshaw* 93/262; and only present on part of *Wood* 173.

Peter spells the epithet as *actinopteroides*; U.K.W.F. (1974) has *actiniopteroides*. Chaerle, in his unpublished thesis, p. 73 explains why he believes that the spelling by Peter was a typographical error, and corrects it to '*actiniopteroides*'. I believe he is right, and I have followed his suggestion.

A. majus differs from *actiniopteroides* in the slightly wider rhizome scales, but not enough to keep the two apart; I have studied the types and decided to treat *A. majus* as a synonym. Chaerle in his unpublished thesis: p. 191 treats *A. majus* as a full species, distinguishing the two by stipe length and glandular covering, as well as in ecological preference.

49. **Asplenium goetzei** *Hieron.* in E.J. 28: 343 (1900); Johns, Pterid. trop. East Africa checklist: 63 (1991); Schippers in Fern Gaz. 14, 6: 201 (1993). Type: Tanzania, Morogoro District: Uluguru Mts, Lukwangule plateau, *Goetze* 283 (B!, holo.)

Epiphyte; rhizome creeping, with dense dark red-brown triangular-ovate entire rhizome scales to 6 × 1.3 mm, long-acuminate into a hairtip. Fronds spaced, erect, to 18 cm long. Stipe 4.5–7 cm long, near base with dense scales similar to those of rhizome, higher mixed up with linear-lanceolate scales, ciliate near base. Lamina elongate-triangular, to 11 × 4.5 cm, 2-pinnate to 3-pinnatifid, gradually decrescent towards apex; pinnae in 6–11 pairs, opposite or alternate, elongate-triangular, to 2.5 × 1–1.3 cm; pinnules deeply 3-partite to 3-lobed on lower pinnae, entire higher up, 3–4-dentate at apex, with lanceolate lacerate-dentate scales. Rachis with scales similar to those of stipe (2 kinds). Sori 1–3 per lobe, linear, 1.5–5 mm long; indusium 0.5 mm wide.

TANZANIA. Morogoro District: Uluguru Mts, Lukwangule plateau, Nov. 1898, *Goetze* 283!
DISTR. **T** 6; not known elsewhere
HAB. Epiphyte, among lichens; 2400 m
CONSERVATION NOTES. Widespread; least concern (LC)

NOTE. Close to *A. actiniopteroides* but with enough differences to keep the two separate. The absence/presence of glands in *A. actiniopteroides* is too variable to use, but the differences in size, the presence of two types of scales in *goetzei*, the slight difference in pinnule apex (with pinnule apices more deeply dissected in *actiniopteroides*) warrant separation.

Pichi Sermolli in B.J.B.B. 55: 135 (1985) and Chaerle in his unpublished thesis, page 155, include several specimens from Rwanda and Congo-Kinshasa under this name. I have not seen the cited specimens (*Bamps* 2963a, *Bequaert* 3711, *Christiaensen* 2238, *Van der Veken* 8864).

50. **Asplenium uhligii** *Hieron.* in E.J. 46: 374 (1912); A.V.P.: 30 (1957); Alston, Ferns W.T.A.: 59 (1959); Tardieu in Fl. Camér. 3, Ptérid.: 214, t. 27/1 (1964); Burrows, S. Afr. Ferns: 254, map, figs. (1990); Johns, Pterid. trop. East Africa checklist: 68 (1991); Faden in U.K.W.F. ed. 2: 30 (1994). Type: Tanzania, Kilimanjaro above Moshi, *Uhlig* 116 (B, lecto. not found, chosen by Tardieu in 1964, photo.!; K!, iso.); syntype *Uhlig* 194 (B!, K, syn.)

Epiphytic, epilithic, or rarely terrestrial; rhizome long-creeping, 2–3 mm diameter, with black or dark to mid-brown ovate-deltate entire attenuate rhizome scales 5–9 × 1–2 mm, ending in a hair-tip. Fronds spaced; stipe purplish black, 4–23 cm long, slender and glabrous or with scattered scales similar to those of rhizome (very rarely with a few capitate glands); lamina narrowly ovate to lanceolate, (4–)8–30 × (2–)3–16 cm, 2-pinnatifid to 3-pinnatifid, basal pinnae longest, apical pinnae decrescent, the pinnae angled forward at 45°; pinnae in 6–13 pairs, dull green, wedge-shaped-rhomboid, 1.5–8 × 0.6–3 cm, acroscopic base truncate-cuneate, basiscopic base cuneate, the lower ones petiolulate, the upper ones sessile, ultimate segments narrowly obcuneate, distal margins finely toothed to crenate, the lower teeth often 2–3-fid, apex long-acuminate or acute; glabrous above, beneath with few scattered dark to mid-brown narrowly triangular and slightly lobed scales along veins; rachis with sparse ovate scales. Sori 1–2 per pinnule segment, linear to oval, 2–7(–10 fide protolog) mm long; indusium brownish, entire, 0.4–0.5 mm wide. Fig. 8: 9–11, p. 47.

UGANDA. Toro District: Ruwenzori, Bujuku valley, Bigo, Mar. 1948, *Hedberg* 433! & Freshfield Pass, July 1951, *Osmaston* 3886!; Kigezi District: Muhavura, Oct. 1948, *Hedberg* 2108!
KENYA. Nyandarua/Aberdare Mts, Feb. 1959, *H.D. van Someren* 1072!; Mt Kenya: W face, bamboo zone, Jan. 1922, *Fries & Fries* 1275! & ridge S of Teleki valley, Aug. 1948, *Hedberg* 1833! & Naro Moru route, Jan. 1970, *Wendelberger* 24!
TANZANIA. Kilimanjaro: between Peter's hut and Bismark's hut, July 1956, *Pichi Sermolli* 5118!!; Mt Meru, upper forest to Ash Cone, 1994, *Grimshaw* 94-112!; Mbeya District: Kikando, Oct. 1956, *Richards* 6706!
DISTR. U 2; **K** 3, 4; **T** 2, 7; Togo, Nigeria, Cameroon, Congo-Kinshasha, Malawi, Zimbabwe
HAB. Giant heath zone, Senecio-Hypericum woodland and upper Podocarpus forest; (2400–)2600–4200 m
CONSERVATION NOTES. Widespread; least concern (LC)

SYN. *A. kassneri* Hieron. in E.J. 46: 376 (1911); A.V.P.: 29 (1957); Pic.Serm. in B.J.B.B. 55: 136 (1985); Johns, Pterid. trop. East Africa checklist: 65 (1991), except *Gillett* specimen. Type: Congo-Kinshasha, Ruwenzori, *Kassner* 3114 (B, holo., not found), **syn. nov**. Note: Pichi Sermolli states the BM and K! sheets of this number are not the same as the B sheet; he has determined these two as *A. volkensii*.
    Asplenium sp. E of Johns, Pterid. trop. East Africa checklist: 68 (1991); Faden in U.K.W.F. ed. 2: 30 (1994)

NOTE. I have reduced *A. kassneri* to synonymy. The protologue already indicated the two taxa were close, and with the amount of material available now there are no remaining consistent differences between what is described in the protologue of *A. kassneri* and *A. uhligii*.
    Pichi Sermolli in B.J.B.B. 55: 124 (1985) did not see the type of *A. kassneri* but saw a photo and a fragment, and treated this taxon on the basis of several *Auquier* specimens in P. These have glandular hairs in varying densities – sadly the B holotype of *A. kassneri* was lent to the BM and is now missing. Similarly Chaerle in his unpublished thesis, page 159, treats *kassneri* as a full species, distinct from *A. uhligii* in having the lamina up to 2-pinnate or more finely dissected, not pinnate to pinnate-pinnatisect

51. **Asplenium praegracile** *Rosenst.* in F.R. 6: 177 (1908); Johns, Pterid. trop. East Africa checklist: 66 (1991); Faden in U.K.W.F. ed. 2: 30 (1994). Type: Tanzania, Kilimanjaro, 3000 m, anno 1906, *Daubenberger* s.n. (B!, holo., fragment; P!, iso.)

Terrestrial or low epiphyte; rhizome creeping, 3–8 mm diameter, with dense dark brown appressed shiny narrowly triangular attenuate scales to 4 × 0.4–0.9 mm long, margin narrow and paler. Fronds well-spaced; stipe erect, dark brown to black, 20–58 cm long, sulcate in upper part, when young covered with scales as on rhizome, but soon glabrescent; lamina dull or dark green, herbaceous to subcoriaceous, ovate-lanceolate, 23–45 × 10–23 cm, 3-pinnatifid to 3-pinnatisect, lowermost pinnae pendent and hardly reduced, upper pinnae gradually decrescent; pinnae 12–20 pairs, stalked, alternate, to 11 × 6 cm, acroscopic base ± parallel to rachis, basiscopic base cuneate, apex attenuate; pinnules in 5–8 pairs, asymmetrically lanceolate, to 3 × 1 cm, most pinnules pinnatisect; with ultimate segments to 2 mm wide, uppermost pinnules with dentate margin; with scattered pale brown clathrate narrowly ovate attenuate scales to 1.5 mm with entire margin. Rachis and costa with scattered pale brown clathrate narrowly ovate attenuate scales to 3 mm with entire margin. Sori several per pinnule, 1–2 per lobe, linear, 3–5.5 mm long; indusium membranous, entire, to 0.3 mm wide. Fig. 8: 12–14, p. 47.

UGANDA. Ruwenzori, 'forest 8000 ft', May 1894, *Scott Elliot* 7641! & Kigezi District: Rutenga, June 1951, *H.D. van Someren* 713! & Kanaba Gap, June 1950, *H.D. van Someren* 648!
KENYA. Aberdare/Nyandarua Mts, Kiandogoro track above Tucha, Oct. 1971, *Faden & Faden* 71/880! & Rotundu, Kzazita R., *Luke & Luke* 4790!

TANZANIA. Kilimanjaro, Bismarck Hill, Feb. 1934, *Greenway* 3838! & near Mandara hut, Oct. 1993, *Grimshaw* 93/785! & above Kilimanjaro Timbers, Jan. 1994, *Grimshaw* 94/38!
DISTR. **U** 2; **K** 4; **T** 2; not known elsewhere
HAB. Upper moist forest zone, bamboo zone, Hagenia zone and heath zone; terrestrial especially on logs, moss tufts, steep banks, or low epiphyte; may be locally common; 2400–3100 m
CONSERVATION NOTES. Fairly widespread with a wide habitat and altitude range; Least Concern (LC)

NOTE. The type, a fragment at B, has a few minute capitate glands on the rachis. *Hemp* 1420 from Kilimanjaro is similar (though sterile); this has a few glands on the lower surface of the lamina as well. No other collections seem to have these capitate glands.
   A specimen from Kenya, **K** 6, Chyulu Hills, May 1938, *H.D. van Someren* 79! looks like it might be this taxon but lacks both rhizome and stipe.

52. **Asplenium mildbraedii** *Hieron.* in Z.A.E. 2: 21 (1910); Pic. Serm. in B.J.B.B. 55: 145 (1985); Johns, Pterid. trop. East Africa checklist: 65 (1991). Type: Congo-Kinshasa, Ninagongo, *Mildbraed* 1369 (B!, syn.); Rwanda, Kahama, *Mildbraed* 1779 (B!, syn.); Tanzania, Usambara, near Magamba above Kwai, *Engler* 1288 (B!, syn.)

Rhizome short-creeping, with two types of rhizome scales plus intermediate ones (HB: no, all the same); dark brown, ovate to narrowly triangular, acuminate, ending in hair-tip, 1–4 × 0.5 mm, from cordate or peltate base, with paler margin. Leaves spaced, 50–80 cm long, 3-pinnatisect. Stipe black, 25–33 cm long, to 2 mm thick, glabrous or nearly so. Lamina oblong in outline, 25–40 × 9–12 cm, apex decrescent; pinnae 15–20, 2-pinnatisect, deltoid-ovate or deltoid-oblong, to 7.5 × 3.5–4 cm; pinnules trapezoid-rhombic, to 2 × 1–1.3 cm, base cuneate, the ultimate segments spatulate with (sub-)truncate and denticulate apex; scaly when young with scales similar to those of rhizome. Rachis with ovate (Kivu) or thread-like and lobed (Ninagongo, Usambara) scales. Sori lateral on veins, 2–5 mm long, not reaching margin; indusium entire, to 0.4 mm wide.

TANZANIA. Lushoto District: Usambara, near Magamba above Kwai, Oct. 1902, *Engler* 1288!
DISTR. **T** 3; Congo-Kinshasa, Rwanda
HAB. Montane or bamboo forest, alpine zone; 2300–3000 m
CONSERVATION NOTES. Fairly widespread with a wide habitat and altitude range; Least Concern (LC)

NOTE. This is close to *A. praegracile* but differs in pinnule shape, which is narrow in *praegracile*, spatulate in *mildbraedii*; and apical lobes are thin and deep in *praegracile*, spatulate and rather shallow in *mildbraedii*.
   The protologue only likens it to *A. furcatum* Thunb. var. *tripinnatum* Baker (± *aethiopicum*) and says it is very like *A. goetzei* var. *major* (now a synonym of *actiniopteroides*) "but with much smaller rhizome scales."
   Chaerle in his unpublished thesis, p. 200, treats it as a full species; and keys it out as having fronds up to 2-pinnate-pinnatisect, as opposed to *praegracile* which he has as up to 3-pinnate or more finely dissected.
   Chaerle adds as syntypes *Libusch* s.n. from Lutindi, and *von Goetzen* 43 from Ninagongo; but Hieronymus only placed these incomplete/juvenile specimens tentatively in this taxon.

53. **Asplenium volkensii** *Hieron.* in P.O.A. C: 83 (1895). Type: Tanzania, Kifinika, *Volkens* 1328 (B, holo. not seen, out on loan, photo! at K)

Terrestrial; rhizome long-creeping, to 5 mm diameter, with dark or mid-brown lanceolate subentire scales to 5 × 1.5 mm long, colour uniform. Fronds widely spaced, 50–85 cm high; stipe 23–32 cm long, with shiny brown triangular-ovate minutely toothed acute scales to 3 mm. Lamina dark green, lanceolate in outline, 32–44 × 18–20 cm, acuminate, 3-pinnatisect to 3-pinnatipartite, lower pinnae very slightly reduced, apical pinna unclear from type (decrescent I think); pinnae in

20–25 pairs, opposite or alternate, triangular-lanceolate in outline, 10–13 × 3–4 cm, prolonged into a pinnatifid-lobed apex; pinnules obliquely ovate or ovate-oblong, 5–7-lobed or -pinnatisect in lowermost pinnules, with basiscopic base dimidiate-cuneate, acroscopic base subtruncate to cuneate with enlarged lobe, apices irregularly subincised dentate-truncate, veins flabellate, subglabrous above, beneath with sparse shiny brown triangular-ovate scales to 1 mm, mainly on veins; rachis and pinnae rachis with scales similar to stipe. Sori mostly along the pinnule centre and a few in the basal lobes, linear-oblong, 2–5 mm long; indusium membranous, entire, ± 0.4 mm wide.

TANZANIA. Kilimanjaro, Kibosho, 1907, *Daubenberger* 62! & Mandara Hut, Feb. 1986, *Schippers* 1202! & above Mandara Hut, Oct. 1993, *Grimshaw* 93/924!
DISTR. **T** 2; not known elsewhere
HAB. Moist forest, *Hagenia* woodland, giant heath zone; 2150–2800 m
CONSERVATION NOTES. Due to the fairly wide habitat and altitude range, this species is assessed as least Concern (LC).

NOTE. All material cited under this name in Pic.Serm. in B.J.B.B. 55: 153 (1985) & Johns, Pterid. trop. East Africa checklist: 68 (1991) is not this taxon. Cited by Schippers in Fern Gaz. 14, 6: 200 (1993) as occurring in Pare and W Usambara – I have seen no material from there.

54. **Asplenium lividum** *Kuhn* in Linnaea, 36: 100 (1869), as *Asplenum*; Tardieu in Mém. I.F.A.N. 28: 195, t. 38/5 (1953); Alston, Ferns W.T.A.: 59 (1959); Tardieu, Fl. Camér. 3, Ptérid.: 210, t. 33/5 (1964); Schelpe, F.Z. Pteridophyta: 181 (1970); Burrows, S. Afr. Ferns: 248, map, figs. (1990); Johns, Pterid. trop. East Africa checklist: 65 (1991); Faden in U.K.W.F. ed. 2: 30 (1994). Type: Venezuela, Tovar, *Fendler* 156 (B!, holo., photo.!)

Epiphytic, less often on wet rocks, rarely terrestrial; rhizome erect or creeping, 3–7 mm diameter, and with mid-brown clathrate lanceolate entire sharply acute rhizome scales 1–3 × 0.2 mm. Fronds tufted or slightly spaced, arching or hanging, firmly herbaceous, 15–65 cm long, not proliferous; stipe matt brown or green above and black beneath, often darker towards the base, 6–20 cm long, with scattered scales similar to those on the rhizome, otherwise glabrous at maturity; lamina narrowly to very narrowly oblong, 13–45 × 4–9(–16) cm, 2-pinnatisect to weakly 3-pinnatifid, lowest pinnae not reduced, apical pinnae decrescent, apical pinna linear and much-lobed; pinnae in 15–21 pairs, subopposite to alternate, unequally rhombic-attenuate, 2–7(–12) × 0.5–2.5(–3) cm, petiolate, very unequally cuneate at the base, the basal acroscopic lobe often oblong and larger than the rest, deeply pinnatifid into 4–9 pairs of narrowly ob-cuneiform to linear truncate-denticulate or truncate-serrate lobes, the lobes of the pinnae ± equal except for the basal acroscopic lobe; glabrous on both surfaces; veins flabellate; rachis matt brown, glabrous at maturity. Sori 1–3 per lobe but also characteristically on the wing along the costa, linear, 1–7 mm long; indusium linear, membranous, entire, 0.3–0.5 mm wide. Fig. 9: 1–2, p. 54.

UGANDA. Toro District: 3 km N of Kichwamba, Sep. 1969, *Faden et al.* 69/1249!; Ankole District: S Kasyoha-Kitomi forest, Kamukaaga, Oct. 1998, *Hafashimana* 672!; Mengo District: Namiryango [Namilyango], Aug. 1915, *Dummer* 2460!
KENYA. Nandi District: Kaimosi, SW of Yala R. bridge, Apr. 1965, *Gillett* 16698!; Thika District: below Chania Falls, Nov. 1968, *Faden* 68/834!; North Kavirondo District: Kakamega forest, NE of Forest Station, Nov. 1969, *Faden et al.* 69/2006!
TANZANIA. Pare District: S Pare Mts, Feb. 1915, *Peter* 9025!; Morogoro District: Nguru Mts, NW Mkobwe, Mar. 1953, *Drummond & Hemsley* 1882!; Iringa District: Mufindi, Lupeme tea estate, May 1968, *Renvoize & Abdallah* 1939!
DISTR. **U** 2, 4; **K** 3–6; **T** 3, 6, 7; Sierra Leone, Cameroon, Angola, Malawi, Mozambique, Zimbabwe, South Africa; tropical America
HAB. Moist forest; 1000–2550 m
CONSERVATION NOTES. Widespread; least concern (LC)

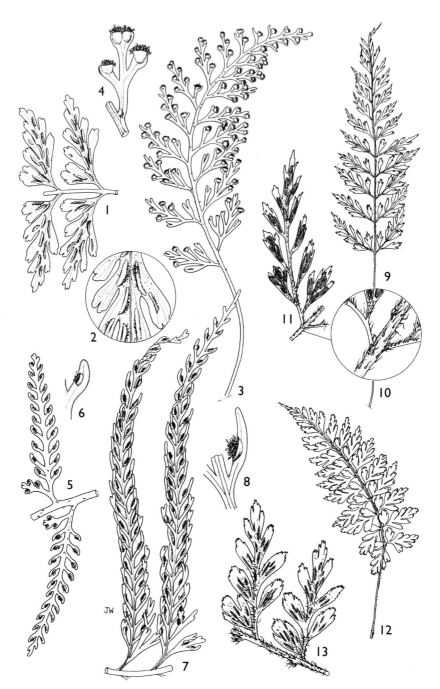

Fig. 9. *ASPLENIUM LIVIDUM* — **1**, pinnae × 1; **2**, sori × 2; *ASPLENIUM THECIFERUM* — **3**, frond × ²/₃; **4**, sori × 2. *ASPLENIUM RUTIFOLIUM* — **5**, pinnae × 1; **6**, sori × 3. *ASPLENIUM LOXOSCAPHOIDES* — **7**, pinnae × 1; **8**, sori × 3. *ASPLENIUM AETHIOPICUM* — **9**, frond × ¹/₃; **10**, detail scales × 2; **11**, pinna × 1. *ASPLENIUM BUETTNERI* var. *BUETTNERI* — **12**, frond × ¹/₃; **13**, pinnae × ²/₃. 1, 2, from *Hafashimana* 210; 3, 4, from *Kibui* 50; 5, 6, from *Grimshaw* 93/582; 7, 8, from *Vesey Fitzgerald* 5059; 9–11, from *Grimshaw* 94/590; 12, 13, from *Hepper & Jaeger* 7023. Drawn by Juliet Williamson.

NOTE. The fronds resemble narrow, weakly 2-pinnatifid fronds of *A. aethiopicum* in size and shape of the lamina, but the taxon differs in smaller rhizome scales; the fronds are also narrower, and the sori are arranged in a different way..
  Gilbert 3467 from **T** 2, Kilimanjaro, Ngare Nairobi north river, 2550 m, Aug. 1969, differs in the many glandular hairs on rachis and upper lamina surface.

55. **Asplenium theciferum** (*Kunth*) *Mett.* in Ann. Sci. Nat., Sér. 5, 2: 227 (1864). Type: Venezuela, 'in temperatis Provinciae', *Humbold & Bonpland* s.n. (P, holo.)

Epiphyte, rarely lithophyte; rhizome erect, 5–12 mm diameter, with dark brown lanceolate attenuate rhizome scales 3–7 mm long, with a paler margin with hair-like lobes. Fronds tufted, somewhat fleshy, erect to arching, to 35 cm long, not proliferous. Stipe greyish-green when dry, 3–18 cm long, very narrowly winged, with occasional brown lanceolate acuminate scales 1–2.5 mm long. Lamina dull green, generally oblong-elliptic to lanceolate, 9–26 × 2.5–7 cm, 2-pinnate to 3-pinnatisect, with basal pinnae hardly reduced. Pinnae 10–17 pairs, alternate, angled upwards, green, oblong to narrowly oblong-obtuse in outline, up to 4 × 1.7 cm, deeply pinnatifid with obliquely linear or spatulate lobes, the basal acroscopic lobe usually 2-fid, glabrous or with sparse scales like those on the stipe, to 1.5 mm long. Rachis grey-green when dried, narrowly winged, with occasional dark brown lanceolate scales to 3 mm. Sori solitary at or just below the apex of the lobe, set off-centre, oblong to broadly oblong, up to 2 mm long and wide, usually subtended on one side by a large or small triangular area of the lamina; indusium oblong to broadly oblong, membranous, entire, to 1 mm wide. Fig. 9: 3–4, p. 54.

SYN. *Davallia thecifera* Kunth, Nov. Gen. Sp. Pl. 1: 23 (1816)

  var. **concinnum** (*Schrad.*) *Schelpe* in Bol. Soc Brot., Sér. 2, 41: 210 (1967); Schelpe, F.Z. Pteridophyta: 188, t. 54d (1970); Burrows, S. Afr. Ferns: 238, map, figs. (1990); Johns, Pterid. trop. East Africa checklist: 67 (1991); Faden in U.K.W.F. ed. 2: 29, t. 173 (1994). Type: South Africa, Cape Province, ?near Grahamstown, *Hesse* s.n. (?LE, holo.)

UGANDA. Toro District: Kibale Forest National Park, May 1997, *Hafashimana* 178!; Ankole District: Buhweju, Nyagoma-Rugongo, Feb. 1990, *Rwaburindore* 2945!; Mt Elgon, May 1923, *Snowden* 791!
KENYA. Northern Frontier District: Mt Nyiru, Mar. 1995, *Bytebier et al.* 74!; N Kavirondo District: Kakamega forest, Mar. 1972, *Kokwaro* 3090!; Teita District: Chawia forest, May 1985, *NMK Taita Hills Expedition* 826!
TANZANIA. Lushoto District: Kwamkoro Forest Reserve, June 1986, *Ruffo & Mmari* 2167!; Mpwapwa District: N Mpwapwa, Aug. 1930, *Greenway* 2444!; Iringa District: Mufindi, May 1968, *Renvoize & Abdallah* 2029!
DISTR. **U** 1–3; **K** 1–7; **T** 2–7; Congo-Kinshasa, Rwanda, Burundi, Angola, Malawi, Mozambique, Zambia, Zimbabwe, South Africa; Madagascar, Comoro Is.
HAB. Low- to high-level epiphyte in forest, occasionally on rocks, sometimes occurring in woodland or thicket; common and widespread; (350–)800–2750(–3150) m
CONSERVATION NOTES. Widespread; least concern (LC)

SYN. *Davallia concinna* Schrad. in Gött. Gel. Anz. 1818: 918 (1818)
  Loxoscaphe concinnum (Schrad.) Moore in Hook. Journ. Bot. 5: 227 (1853)
  Asplenium concinnum (Schrad.) Kuhn, Fil. Afr.: 99 (1868)
  A. theciferum sensu Sim, Ferns S. Afr. ed. 2: 171, t. 72 (1915), *non* (Kunth) Mett. (1864)
  Loxoscaphe theciferum (H.B.K.) T.Moore var. *concinna* (Schrad.) C.Chr. in Dansk Bot. Ark. 7: 104 (1932), as *concinnum*; Pic. Serm. in B.J.B.B. 55: 124 (1985)

NOTE. The typical variety is from South America.
  Plants with ovate-deltoid laminas (from **U** 2, Kigezi, and **K** 5, Kakamega) have been identified as being close to, or identical with, *A. cornutum* Alston in Bol. Soc. Brot. ser. 2, 30: 8 (1956) & Ferns W.T.A.: 59 (1959); Johns, Pterid. trop. East Africa checklist: 62 (1991). I see no real difference and believe the Alston name to be synonymous with *A. theciferum* sensu lato. As the type of *A. cornutum* is from Cameroon I will not make a formal decision regarding synonymy.

56. **Asplenium rutifolium** (*Bergius*) *Kunze* in Linnaea, 10: 521 (1836), as *rutaefolium*; Schelpe, F.Z. Pteridophyta: 185 (1970); Burrows, S. Afr. Ferns: 236, map, figs. (1990); Johns, Pterid. trop. East Africa checklist: 67 (1991). Type: South Africa, Cape of Good Hope, *Thunberg* s.n. (SBT, holo.)

Lithophyte, epiphyte or terrestrial; rhizome erect, to 50 mm long and to 25 mm diameter, with dark brown lanceolate to narrowly ovate acute rhizome scales 4–9 × 0.8–1.2(–2.2) mm with or without narrow pale borders, subentire or sometimes with a few hair-like lobes. Fronds tufted (once described as rhizome shortly creeping, fronds shortly spaced, *Faden* 69/327), usually stiffly erect, not proliferous. Stipe brown abaxially, green adaxially, 3–20 cm long, either with scattered scales to 3 × 1.3 mm with dark brown midpart and broad pale margins or glabrous except for scattered minute dark brown ovate acuminate scales, becoming glabrous with age. Lamina ovate to narrowly oblong in outline, 7–40 × 3–10(–12) cm, thinly to thickly coriaceous, 2-pinnatisect to 3-pinnatisect, with basal pinnae hardly or not reduced, apex gradually decrescent; pinnae 13–28 pairs, oblong and attenuate or obtuse, the largest 3–7 × 0–7–1.3(–1.7) cm, shortly stalked, deeply pinnatifid into mostly linear or very narrowly spatulate segments set at 45°, 3.5–8 × 0.6–1.2 mm (up to 2 mm across sori) but with the acroscopic basal segment often 2–9-lobed, often with some of the basal segments 2-fid to 2-lobed, apices obtuse, basiscopic base sub-dimidiate, glabrous on both surfaces except for occasional minute dark narrowly ovate scales. Rachis narrowly winged or ridged laterally, pale brown when dry with occasional minute dark brown ovate scales. Sori 1 per pinna lobe, ellipsoid, borne halfway along the length of the lobe or slightly higher and reaching from the costule to or beyond the margin, facing towards the pinna apex, 1.3–3.5 mm in our area, elsewhere 0.8–5 mm long; indusium oblong, membranous, entire, to 0.8 mm wide. Fig. 9: 5–6, p. 54.

UGANDA. Ankole District: Ruizi R., Mar. 1951, *Jarrett* 406!; Mt Elgon, below Bulambuli, Dec. 1957, *Allen* 3671!
KENYA. Nothern Frontier District: Ndoto Mts, between Nkurnit and Manmanet ridge, Oct. 1995, *Bytebier & Kirika* 42!; Nairobi, no exact locality, May 1971, *Mwangangi & Mukenya* 1599!; Teita District: road to Bura Mission, *Faden et al.* 69/327!
TANZANIA. Musoma District: Wogakuria guard post, Dec. 1964, *Greenway & Turner* 11798!; Lushoto District: Baga Forest Reserve, May 1987, *Kisena* 477!; Iringa District: Mufindi, Luiga tea estate, May 1968, *Renvoize & Abdallah* 2057!
DISTR. **U** 2, 3; **K** 1, 3–7; **T** 1–4, 6–8; Malawi, Mozambique, Zimbabwe, South Africa; Madagascar, Mascarene Is; Yemen
HAB. Epiphyte (high, medium and low), lithophyte or terrestrial, in moist and dry evergreen forest, riverine forest; 750–2100(–2300) m
CONSERVATION NOTES. Widespread; least concern (LC)

SYN. *Lonchitis bipinnata* Forssk., Fl. Aegypt.-Arab.: CXXIV, 184 (1775)
    Adiantum achilleifolium Lam., Encycl. Méth., Bot. 1: 43 (1783). Type: South Africa, Cape of Good Hope (no collector indicated)
    Caenopteris rutifolium Bergius in Act. Petropol. 1782, 2: 249, t. 7 fig. 2 (1786), as *rutaefolia*
    Darea disticha Kaulf., Enum. Fil.: 180 (1824), *nom. illegit.* Type as for *Lonchitis bipinnata*
    Caenopteris disticha Spreng. in L., Syst. Nat., ed. 16, 4: 91 (1827)
    Asplenium achilleifolium (Lam.) C.Chr., Ind. Fil.: 99 (1905), *non* Liebm. (1849)
    A. distichum (Kaulf.) Salomon, Nomencl. Gefässkrypt.: 84 (1883)
    A. bipinnatum (Forssk.) C.Chr, Index filic.: 99 (1905) & apud Hieron., Z.A.E.: 11 (1910). Type: Yemen, Bolghose, *Forsskål* 810 (C, holo., B! (fragm.), iso.), *non A. bipinnatum* Roxb. (1844) *nec* Hil. (1888)
    A. linearilobum Peter in F.D.O.-A.: 80 (1929) & Descr.: 8, t. 2.7–2.8 (1929); Johns, Pterid. trop. East Africa checklist: 65 (1991). Type: Tanzania, Lushoto District: W Usambara, Manolo–Mtai, *Peter* 4196c (B, holo., not found). Note: Peter also says this is *A. borbonicum* Jacq., Collect.: 3 (1789), *non* Hook.; though he calls 4196 'specimen authenticum', which is his term for type!
    A. rutifolium (Berg.) Kunze var. *bipinnatum* (Forsk.) Schelpe in Journ. S. Afr. Bot.30: 194 (1964); Schelpe, F.Z. Pteridophyta: 180, t. 54c (1970); Johns, Pterid. trop. East Africa checklist: 67 (1991)

A. *rutifolium* (Berg.) Kunze var. *linearilobum* (Peter) Schelpe ined.?; Johns, Pterid. trop. East Africa checklist: 67 (1991). Nomen nudum? used on quite a few herbarium sheets. Note: in F.Z. Schelpe puts *linearilobum* Peter as a synonym of his *A. rutifolium* var. *bipinnatum*.

A. *strangeanum* Pic.Serm. in Webbia 32 (1): 83, t. 4 (1977) & in B.J.B.B. 55: 152 (1985); Schippers in Fern Gaz. 14, 6: 204 (1993); Faden in U.K.W.F. ed. 2: 29, t. 173 (1994). Type: Kenya, Kiambu District: Kiambu, Chania R., *Strange* 59 (herb. Pic.Serm., holo.), **syn. nov.**

NOTE. Burrows has re-united var. *bipinnatum* with the typical variety, due to the presence of intermediates. I am happy to follow him.

A. *strangeanum* agrees completely with *A. rutifolium* – as does the illustration with the protologue. The only differences seem to be slightly longer rhizome scales. The author did not compare it to *rutifolium*, but only to *A. borbonicum* – which does not occur in East Africa.

H.D. van Someren 102 from Kenya **K** 7, Shimba Hills, is sterile but looks like this species – it would be a low altitude, but the specific altitude is not given on the label.

There are a whole range of problem specimens falling in between *rutifolium* and *loxoscaphoides*, with a worrying variability of rhizome scales – narrow ones about 1 mm wide, with a dark brown middle part and a thin pale brown margin are usual for *rutifolium*; wider ones, up to 2.5 mm wide, with a narrow dark brown central part and a wide pale brown margin, occur in specimens from northern Tanzania and Kenya, e.g. *Beesley* 10! from Kilimanjaro and *Archbold* 970! from the same area. This form of rhizome scale is often, but not always, linked to longer stipes, of up to 25 cm. Confusingly, in this same area 'normal' *rutifolium* specimens with the narrower scales are also common. It is possible that specimens with narrow pale margins and a rather broad dark centre are linked to a moist forest habitat.

I believe there are some differences, as per the key (rhizome scales being the most clear; plus some relative sizes) but the two taxa are close.

57. **Asplenium loxoscaphoides** *Baker* in Trans. Linn. Soc. 2: 354 (1887); Johns, Pterid. trop. East Africa checklist: 65 (1991); Faden in U.K.W.F. ed. 2: 29, t. 173 (1994). Type: Tanzania, Kilimanjaro, *Johnston* 43 (K!, holo.)

Terrestrial, epiphyte or rarely lithophyte; rhizome erect, fleshy, to 60 mm long and 8–50 mm diameter, with pale to mid-brown ovate acute subentire rhizome scales to 9–14 × 3–3.5 mm. Fronds tufted, erect to arching, not proliferous, thinly to thickly coriaceous. Stipe pale brown to greyish-green when dry (green to black when fresh), 15–48 cm long, glabrous except for some scales near base. Lamina dark green, narrowly ovate in outline, 32–80 × 13–26 cm, 2-pinnatisect to 3-pinnatifid on the lowermost pinnules, with basal pinnae reduced (rarely not reduced), apex gradually decrescent; pinnae opposite or alternate, 23–44 pairs, oblong and attenuate, the largest (6–)8–13(–16) × 1–2.2 cm or up to 4 cm wide with much-lobed basal pinnules, petiolate, deeply pinnatifid into linear or very narrowly spatulate segments set at 45°, 1–2 mm wide, most segments entire or some with bifid apex or many segments bifid, basal acroscopic segment usually more lobed or enlarged to an up to 2.5 cm long deeply lobed pinnule (rarely basiscopic segment also more lobed), apices acute or obtuse; more basal pinnae often shorter and with more widely winged costa and so appearing more pinnatipartite than the pinnatisect upper pinnae; with scattered scales similar to those on rachis. Rachis black and green when fresh, pale brown when dry, with scattered dark brown narrowly lanceolate scales to 3 mm long with hair-like lobes, sometimes becoming glabrous. Sori 1 per pinna lobe, ellipsoid, borne halfway along the length of the lobe but almost on the margin, facing towards the pinna apex, 1.5–5(–6) mm long; indusium very narrowly oblong, membranous, entire, to 1 mm wide. Fig. 9: 7–8, p. 54.

KENYA. Mt Elgon, E slope above Tweedie's saw-mill, Feb. 1948, *Hedberg* 116!; Meru District: Ithangune, near Katheri, June 1969, *Faden et al.* 69/780!; Masai District: Chyulu Hills, Oct. 1969, *Gillett & Kariuki* 18857!
TANZANIA. Kilimanjaro, above Mandara Hut, Oct. 1993, *Grimshaw* 93/860!; Masai District: Embagai crater rim, Dec. 1956, *Greenway* 9138!; Morogoro District: Uluguru Mts, Bondwa SF., Sep. 1970, *Faden et al.* 70/638!
DISTR. **K** 1, 3–6; **T** 2, 6, 7; north Malawi

Hab. Moist montane forest, bamboo zone, cedar forest, Hagenia woodland, giant heath zone; terrestrial, low to medium epiphyte, or in rock crevices; may be locally common; 1850–3100(–3650) m
Conservation notes. Fairly widespread in a series of habitats and often locally common; Least Concern (LC)

Note. Baker in his protologue said the taxon was close to *A. rutifolium*; the characters he uses in the description are similar to those of *rutifolium*, except for 'lower pinnae 10–12.5 cm long, 10–12 mm wide' – which is larger than in *rutifolium*. The type (and only specimen Baker saw) does not have a stipe & rhizome. Baker did not mention how he thought the plant differed from *rutifolium*.
   I believe there are some differences, as per the key (rhizome scales being the most clear; plus some relative sizes) but the two taxa are close, and there are some vexing overlap characters/specimens (see note under *A. rutifolium*).
Asplenium centrafricanum Pic.Serm. in Webbia 27: 436 (1972 publ. 1973) from Burundi, Zaire is probably the same; the differences enumerated in the protologue do not sound very different to me.

58. **Asplenium aethiopicum** (*Burm.f.*) *Becherer* in Candollea, 6: 23 (1935); A.V.P.: 29 (1953); Alston, Ferns W.T.A.: 59 (1959); Schelpe, F.Z. Pteridophyta: 181 (1970); Burrows, S. Afr. Ferns: 246, map, figs. (1990); Johns, Pterid. trop. East Africa checklist: 61 (1991); Thulin, Fl. Somal. 1: 13 (1993); Faden in U.K.W.F. ed. 2: 30, t. 173 (1994). Type: South Africa, Cape, herb. *Burmann* s.n. (G, holo. fide Burrows; but no type indicated in protologue, so this must be lecto.)

Epiphyte, lithophyte or occasionally terrestrial; rhizome erect or shortly creeping, occasionally long-creeping, up to 7 mm diameter, with dark brown to blackish subulate to narrowly triangular (sub-)entire attenuate clathrate scales 3–7 × 0.3–0.6 mm, with very narrow paler margin, ending in a hair-point. Frond tufted to shortly spaced, usually arching, not proliferous, firmly herbaceous to thinly coriaceous. Stipe dark brown to blackish, 4–40 cm long, covered at first with a mixture of scales similar to those on rhizome and mid-brown basally ovate, attenuate clathrate often rather hair-like scales to 3 mm with margin entire or with a few hair-like outgrowths, becoming subglabrous with age except near base. Lamina ovate to lanceolate in outline, 8–60 × 2.5–16 cm, 2-pinnate to 3-pinnatifid or even 3-pinnatisect on lowermost pinnules, with the lower pinnae not reduced or occasionally slightly reduced; apical pinnae gradually decrescent. Pinnae in 7–25 pairs, dark to olive green, deltoid to narrowly lanceolate, and pinnatifid to 2-pinnatifid, 1.5–12 × 0.7–4 cm, base stalked, unequally cuneate, divided into narrowly obcuneate to narrowly oblong segments, pinnules in general irregularly incised and serrate or crenate at their apices, scaly to glabrous beneath, sparsely scaly to glabrous on the upper surface. Rachis dark brown to blackish, green above and distally, covered with mostly hair-like scales but with some lanceolate scales, similar to those on the stipe, becoming subglabrous with age. Sori closely packed along the veins, linear, 2.5–10 mm long, when sporulating may obscure lower surface of pinnule; indusium very narrowly linear, membranous, subentire, 0.3–0.5 mm wide. Fig. 9: 9–11, p. 54.

Uganda. Karamoja District: Mt Kadam, Obda peak, Apr. 1953, *G.H.S.* Wood 947!; Ankole District: Buhweju, Rwakondo, Jan. 1990, *Rwaburindore* 2935!; Mt Elgon, Suam ridge, June 1997, *Wesche* 1412!
Kenya. Northern Frontier District: Ndoto Mts, Manmanet ridge, Oct. 1995, *Bytebier & Kirika* 34!; Nanyuki District: Ngare Ndare forest, Apr. 1985, *Beentje* 2098!; Teita District: Taita Hills, Mbololo, May 2000, *Wakanene & Mwangangi* 260!
Tanzania. Meru District: Olmotonyi Forestry Institute, Mar. 1983, *Mtui* 146!; Lushoto District: Shume–Maramba, May. 1987, *Kisena* 632!; Njombe District: Livingstone Mts, Malaba Mt, Jan. 1991, *Gereau & Kayombo* 3606!
Distr. U 1–4; K 1–7; T 1–4, 6–8; throughout tropical Africa and to South Africa; Indian Ocean islands

HAB. Low to high epiphyte, in rock crevices or on mossy rocks, much less often terrestrial (at least in our area); moist or dry forest, wooded ravines, rocky woodland, bushland or bushed grassland, bamboo zone, moorland; 1150–3000(–3700) m

CONSERVATION NOTES. Widespread; least concern (LC)

SYN. *Trichomanes aethiopicum* Burm. f., Fl. Cap. Prodr. in Fl. Ind.: 32 (err. 28) (1768)

    Asplenium adiantoides Lam., Encycl. Méth. Bot. 2: 309 (1786), *non* (L.) C.Chr. (1905). Type: South Africa, *Sonnerat* s.n. (P, holo., not found)

    A. falsum Retz., Obs. Bot. 6: 38 (1791). Type: South Africa, False Bay, no collector indicated

    A. furcatum Thunb., Prodr. Pl. Cap.: 172 (1800). Type: South Africa, no locality or collector indicated

    Tarachia furcata (Thunb.) C.Presl, Epim. Bot.: 80 (1851) reimpr. in Abh. Königl. Böhm. Ges. Wiss., ser. 5, 6: 440 (1851)

    Asplenium albersii Hieron. in E.J. 46: 380 (1912); Johns, Pterid. trop. East Africa checklist: 61 (1991); Schippers in Fern Gaz. 14, 6: 198 (1993). Type: Tanzania, Usambara, Kwai, *Albers* 289 (B!, lecto., chosen by Pichi-Sermolli 1974 according to detslip on B sheet); Kwai, Gomba Mt, *Buchwald* 295 (B!, K! [see note], syn.); Lutindi, anno 1902, *Liebusch* s.n. (B!, syn.), **syn. nov.**

    A. praemorsum sensu Sim, Ferns S. Afr. ed. 2: 163, t. 65, 66 (1915), *non* Sw. (1788)

NOTE. An extremely variable species, in both size, leaf dissection and pinna shape. It seems likely there is more than one taxon involved, and/or hybridization; but for practical reasons, it is treated here as a single ('sensu lato') taxon. Plants from above 3000 m altitude have shorter stipes and pinnae, and rarely exceed 20 cm; plants growing on rocks are also usually rather small, and possibly plants exposed to full sun in such situations have the fronds less dissected.

    Braithwaite in Bot. J. Linn. Soc. 93: 343–378 describes four subspecies from South Africa, with a few references to East African material. The taxa are keyed out by agamospermy (subsp. *filare* A.F.Braithw.), leaf dissection and spore characters, and include subsp. *tripinnatum* (Baker) A.F.Braithw. and subsp. *dodecaploideum* A.F.Braithw. Chaerle, in his unpublished thesis on the afromontane Aspleniums, follows Braithwaite; I am unable to write a working key to these taxa and so prefer to keep this taxon as 'sensu lato' (if not latissimo) for the time being.

    I have added *A. albersii* to the synonymy; with some reluctance, as it presents a form that is reasonably distinct in the narrowly obovate segments. This form however does not key out, all its characters being within the range of variation of *A. aethiopicum*; nor is there a geographical or ecological gap. The *A. albersii* form occurs in **K** 7; **T** 3, 6, 7; and is found in moist forest between 900–2000 m. Two of its types, *Buchwald* 295 and *Libusch* s.n., are *A. aethiopicum* sensu stricto, and only the third represents the *albersii* facies. An added problem is that this brings *A. buettneri* even closer to the main *aethiopicum* group... *Asplenium albersii* Hieron. var. *eickii* Hieron. is a synonym of *A. linckii*.

    The most closely related other species are the following:

    A. praegracile, distinguished by more finely dissected lamina and the creeping rhizome

    A. blastophorum, distinguished by proliferous fronds

    A. buettneri distinguished by its obovate, rounded pinnules

    A. lividum distinguished by the sori in the wings along the rachis

    A. uhligii and *A. volkensii* distinguished by their widely spaced fronds on a creeping rhizome

    A plant from Kenya, **K** 3, Twins Hills [not traced], 00°15'S 36°35'E [which makes this N Aberdares], rock crevices at 3300 m, Sep. 1976, *Schippers* K369, has few and short pinnae with narrow lobes (about 1 mm wide) but the scale mixture of this taxon. I believe it to be an aberrant high-altitude form, but it looks strange enough to mention it as something to look out for.

59. **Asplenium buettneri** *Hieron.* in Z.A.E. 2: 23, t. 2 fig. 2 (1910); Alston, Ferns W.T.A.: 59 (1959); Schelpe, F.Z. Pteridophyta: 182 (1970); Pic.Serm. in B.J.B.B. 55: 127 (1985); Johns, Pterid. trop. East Africa checklist: 62 (1991); Faden in U.K.W.F. ed. 2: 30 (1994). Type: Togo, Misa Hills, *Baumann* 42 (B!, lecto., P!, iso.), chosen by Tardieu, Fl. Cam. 3: 215 (1964)

    Lithophytic or terrestrial; rhizome creeping or erect, up to 4 mm diameter, with subulate dark brown entire attenuate rhizome scales 2–5 × 0.3–0.6 mm, ending in a hair-point and with marginal cells slightly paler. Fronds spaced to tufted, erect, not proliferous, thinly coriaceous to thinly herbaceous. Stipe dull brown to black at least at the base, paler above in younger specimens, 6–37 cm long, with scales similar to those of rhizome, higher up with mid-brown clathrate subulate attenuate scales to 2 mm with a few thin lobes near the base. Lamina narrowly oblong to triangular-ovate in outline, 8–26(–30) × 4–15(–18) cm, 2-pinnate to 3-pinnatifid or 3-pinnatisect at base; basal pinnae not or hardly reduced. Pinnae dark green above, ovate or rhombic, up to 8 × 4 cm, with obovate segments, the larger segments stalked and up to 23 × 13 mm, sometimes lobed, base asymmetrically cuneate, distal margin rounded and serrate or gradually tapering and serrate, glabrous on both surfaces or with scattered minute scales; basal pinnules largest, apical pinnule also large, ovate and lobed. Rachis black dorsally, green and sulcate ventrally, not winged or sometimes narrowly winged, with clathrate brown lanceolate scales ciliate at their base, up to 1.5 mm long. Sori 5–12 per pinnule, along the veins, linear, 3–13 mm long; indusium linear, membranous, entire, 0.4 mm wide.

var. **buettneri**

Rhizome short-creeping or erect, rhizome scales 2–4 × 0.4–0.6 mm. Fronds tufted. Stipe 7–37 cm long. Lamina 14–26(–30) × 7–15(–18) cm. Rachis not winged. Sori 3–7 mm long. Fig. 9: 12–13, p. 54.

UGANDA. Kigezi District: Bwindi National Park, Kayonza, Feb. 1995, *Poulsen et al.* 732!; Masaka District: E of Lake Kayanja, July 1974, *Katende* 1191!; Mengo District: 3 km E of Entebbe, Sep. 1949, *Dawkins* 375!
KENYA. Northern Frontier District: Mt Kulal near Gatab, Nov. 1978, *Hepper & Jaeger* 7023!; Fort Hall District: Mitubiri, Jan. 1956, *Napper* 490!; Teita District: Vuria hill, May 1985, *NMK Taita Hills Exped.* 316!
TANZANIA. Bukoba District: Minziro Forest, Sep. 1952, *Procter* 104!; Ufipa District: Milepa, Feb. 1950, *Bullock* 2488!; Chunya District: Igila Hill, Kepembawe, Mar. 1965, *Richards* 19811!
DISTR. **U** 2–4; **K** 1, 3, 4, 7; **T** 1, 4, 6, 7; west and central tropical Africa
HAB. In rock crevices, on thin soil over rock or terrestrial within moist or riverine forest; 1100–1500(–2100) m
CONSERVATION NOTES. Widespread; least concern (LC)

NOTE. I am not very happy about the distinction between this taxon and *A. aethiopicum* – it seems to be based only on pinnule shape, which is extrememely variable. The cited sheet from **K** 1 is very close to other specimens named *aethiopicum*, because the lower pinnules are a bit incised at the apex.
   Poulsen et al. 732 from **U** 2, Bwindi National Park, alt. 1400 m, is *buettneri* -but has a creeping rhizome with spaced fronds; rhizome scales are 0.5 mm wide. This seems to be intermediate with var. *hildebrandtii*. *Faden et al.* 70/651 from **T** 6, Uluguru Mts, also seems to be intermediate.

var. **hildebrandtii** *Hieron.*, Z.A.E. 2: 25 (1910); Schippers in Fern Gaz. 14, 6: 199 (1993), as subsp. (lapsus calami?). Type: Kenya, mainland near Mombasa, *Hildebrandt* 1958 (B!, K!, syn.) & Nyika, *Wakefield* s.n. (K!, syn.) & 'Nyika', *Grant* s.n. (B!, syn.); Tanzania, Lushoto District: Amani, *Warnecke* 475 (B!, EA!, P!, syn.)

Rhizome long-creeping, rhizome scales 2–5 × 0.3 mm. Fronds spaced to loosely tufted. Stipe 6–22 cm long. Lamina 8–24 × 4–15 cm. Rachis sometimes narrowly winged. Sori 3–13 mm long.

KENYA. Kitui District: crossing of Tiva R. on Kangondi–Kitui road, May 1971, *Faden & Evans* 71/380!; Kilifi District: Mangea Hill, Apr. 1989, *Robertson & Luke* 5742a!; Kwale District: Shimba Hills, Marere, Mar. 1991, *Luke & Robertson* 2743b!
TANZANIA. Lushoto District: East Usambaras, Sigi Singali, Apr. 1950, *Verdcourt* 166! & Marimba Forest Reserve, Nov. 1986, *Iversen et al.* 86/930!; Lindi District: Rondo Plateau, Rondo Forest Reserve, Feb. 1991, *Bidgood et al.* 1431!
DISTR. **K** 4, 7; **T** 3, 6, 8; Mozambique
HAB. Evergreen or semi-evergreen forest, riverine forest, dense thicket under miombo; 30–500(–1200) m
CONSERVATION NOTES. Widespread; least concern (LC)

SYN. *A. buettneri* sensu Burrows, S. Afr. Ferns: 251, map, figs. (1990)

NOTE. The variety is distinct in the long-creeping rhizome with more widely spaced fronds, as well as the narrower rhizome scales to 0.3 mm wide (instead of 0.4–0.6 mm).

60. **Asplenium varians** *Hook. & Grev.*, Ic. Fil.: t. 172 (1830); Schelpe, F.Z. Pteridophyta: 177 (1970); Johns, Pterid. trop. East Africa checklist: 68 (1991); Faden in U.K.W.F. ed. 2: 29 (1994). Type: Nepal, on rocks, local name 'Dawecow', 1818, *Wallich* s.n. (K!, holo.)

Lithophyte or low epiphyte; rhizome erect, 2–9 mm diameter, with dark brown clathrate-lanceolate attenuate subentire rhizome scales up to 3 mm long. Fronds tufted, erect or arching, not proliferous, herbaceous. Stipe greyish-green when dried, becoming dark brown at the base, 1–13 cm long, with scattered subulate dark brown weakly pseudoserrate scales up to 2 mm long. Lamina narrowly elliptic in outline, 5–23 × 1.5–5 cm, 2-pinnatifid to 2-pinnatisect, basal 2–3 pairs slightly reduced; pinnae up to 10 pairs, to 3 × 1.3 cm, pinnules and pinnule lobes obovate or spatulate, the basal acroscopic lobe usually free, up to 6 mm long, sharply dentate to shallowly incised on the outer margin, glabrous above, subglabrous beneath. Rachis greyish-green when dried with narrow wings, glabrous. Sori 2–5 per lobe, linear to oval, 2–4 mm; indusium linear, almost transparent, erose, 0.3–0.5 mm wide. Fig. 10: 1–2, p. 62.

subsp. **fimbriatum** (*Kunze*) *Schelpe* in Bol. Soc Brot., Sér. 2, 41: 211 (1967); Schelpe, F.Z. Pteridophyta: 177, t. 53e (1970); Burrows, S. Afr. Ferns: 242, map, figs. (1990); Johns, Pterid. trop. East Africa checklist: 68 (1991). TAB. 53 fig. E. Type: South Africa, Natal, between Omfondi & Tugeal Rs., *Gueinzius* s.n. (W, holo., HBG, K!, iso.)

UGANDA. Karamoja District: Mt Kadam [Debasien], July 1949, *H.D. van Someren* 598! & idem, Namojongotyang, Jan. 1936, *Eggeling* 2622!; Toro District: Buranya county, 2 km N of Kichwamba, Sep. 1969, *Faden et al.* 69/144!
KENYA. Turkana District: Murua Nysigar, Feb. 1965, *Newbould* 7228!; Trans Nzoia District: Elgon, July 1949, *H.D. van Someren* 606!; Nairobi, Karura Forest, no date, *Gardner* 972!
TANZANIA. Arusha District: Lake Duluti, May 1965, *Beesley* 139!; Kilimanjaro, forest above Lerangwa, Jan. 1994, *Grimshaw* 94/179! & E of Legumishera Hut, Apr. 1994, *Grimshaw* 94/415!
DISTR. U 1, 2; **K** 2–4; **T** 2; E Congo-Kinshasa, Mozambique, Zimbabwe, South Africa
HAB. Rock crevices in dry evergreen forest or riverine forest, occasionally on tree stumps; 1200–2200 m
CONSERVATION NOTES. Widespread; least concern (LC)

SYN. *Asplenium fimbriatum* Kunze in Linnaea, 18: 117 (1844)

NOTE. The typical variety occurs in Asia.

61. **Asplenium adiantum-nigrum** *L.*, Sp. Pl. 2: 1081 (1753); A.V.P.: 28 (1953); Alston, Ferns W.T.A.: 59 (1959); Burrows, S. Afr. Ferns: 244, map, figs. (1990); Johns, Pterid. trop. East Africa checklist: 61 (1991); Faden in U.K.W.F. ed. 2: 29 (1994). Type: Dodoens, Stirp. Hist. Pempt.: 466 (1616)

Terrestrial or in rock crevices; rhizome shortly creeping, to 4 mm thick, with dark brown clathrate narrowly triangular attenuate rhizome scales to 2 mm long. Fronds tufted; stipe shiny black or dark red-brown, 2–23 cm long, glabrous but for some scales near the very base, swollen at base; lamina deltoid to narrowly ovate in outline, 4–26 × 2–14 cm, 3-pinnatisect, basal pinnae largest, apical pinna narrow and much lobed; pinnae 7–15 pairs, broadly to narrowly ovate, to 8(–12) × 4.5(–5.5) cm, lobes very dissected near base of pinna, margins serrate to dentate, glabrous above, (sub)glabrous beneath; rachis black near base, straw-coloured higher up, glabrous. Sori 2–5 per lobe, tightly clustered along costa and costules and at a slight angle, ovate, 1–2.5 mm long; indusium translucent, 0.5–0.7 mm wide, entire. Fig. 10: 3–4, p. 62.

FIG. 10. *ASPLENIUM VARIANS* — **1**, frond × ²/₃; **2**, pinnae × 1; *ASPLENIUM ADIANTUM- NIGRUM* — **3**, frond × ²/₃; **4**, pinna × 2. *ASPLENIUM ABYSSINICUM* — **5**, frond × ²/₃; **6**, pinna × 1. *ASPLENIUM PUMILUM* — **7**, frond × ²/₃. 1, 2, from *van Someren* 598; 3, from *Cameron* 133; 4, from *Faden* 72/250; 5, from *Thulin & Tidigs* 57; 6, from *E.J. & C. Lugard* 354; 7, from *Hannington* s.n. Drawn by Juliet Williamson.

KENYA. Northern Frontier District: Mt Nyiru, Dec. 1972, *Cameron* 133!; Mt Elgon, crater, May 1948, *Hedberg* 906!; Machakos District: Chyulu Hills eastern edge, Dec. 1996, *Luke & Luke* 4567!

TANZANIA. Mt Kilimanjaro: Lent Valley near Moir hut, Sep. 1993, *Grimshaw* 93/643! & near Sheffield Camp above rain gauge, Aug. 1993, *Grimshaw* 93/500!; Mt Meru, crater, Dec. 1966, *Vesey-Fitzgerald* 5048!

DISTR. **K** 1, 3, 4, 6, 7; **T** 2; Cameroon, Chad, South Africa; Morocco, Algeria; Europe, N America and Hawaii

HAB. Moist lava or rock, roadside and shamba banks, heath scrub, moorland; 1650–4300 m

CONSERVATION NOTES. Widespread; least concern (LC); uncommon in our area

NOTE. In very small plants (those from high altitudes) the sori may obscure the pinna surface.

62. **Asplenium abyssinicum** *Fée*, Gen.: 199 (1850–52); Mett. n. 97 (1850); A.V.P.: 27 (1957); Alston, Ferns W.T.A.: 57 (1959); Johns, Pterid. trop. East Africa checklist: 61 (1991); Faden in U.K.W.F. ed. 2: 29 (1994). Type: Ethiopia, Mt Selki [Silke], *Schimper* 679 (B!, holo.; K!, iso.)

Terrestrial, lithophyte or low/medium epiphyte; rhizome erect, dark brown, with rhizome scales mid- to dark brown, linear and attenuate, to 5.5 × 0.8 mm, margin entire, apex attenuate. Fronds tufted, erect, not proliferous, herbaceous. Stipe dark brown, purplish or black, 3–30 cm long, at the very base with scales similar to those of the rhizome, otherwise glabrous. Lamina membranous, delicate, bright green, ovate to lanceolate in outline, 10–40(–70) × 5–20 cm, 3-pinnate, the lowermost pinnae reduced, decrescent towards the apex. Pinnae mid-green, narrowly triangular in outline, up to 11 × 3.5 cm, unequally cuneate at base, attenuate, divided into ovate or rhombic-ovate pinnules, these lobed to pinnate, the acroscopic base parallel to the rachilla, the basiscopic base attenuate, the ultimate segments obovate, with rounded to obtuse apices, glabrous or with sparse scales to 0.1 mm long. Rachis brown with thin green wings, glabrous or nearly so. Sori 1–2 per segment, inserted on the veins and facing towards the costule, ovoid, 0.5–2 mm long; indusium half-circular, membranous, entire, 0.7–1 mm wide. Fig. 10: 5–6, p. 62.

UGANDA. Karamoja District: Mt Kadam, 1959, *J. Wilson* 761!; Ruwenzori, 7 km above Kilembe, Feb. 1969, *Lock* 69/10!; Mt Elgon, Sasa trail, Feb. 1997, *Wesche* 866!
KENYA. Northern Frontier District: Mt Nyiru, Dec. 1972, *Cameron* 149!; Nakuru District: Eburru Forest Reserve, July 2002, *Luke et al.* 8978!; Mt Kenya, near Urumandi Hut, Jan. 1986, *Beentje* 2629!
TANZANIA. Arusha District: Arusha National Park, Bilo R., Apr. 1968, *Greenway & Kanuri* 13473!; Ufipa District: Mbizi Forest Reserve, Oct. 1987, *Ruffo & Kisena* 2794!; Iringa District: Udzungwa Mt National Park, Sep. 2000, *Luke et al.* 6759!
DISTR. **U** 1–3; **K** 1, 3, 4; **T** 2, 4, 6, 7; Sudan, Ethiopia, Rwanda, Burundi
HAB. Moist forest, bamboo zone, *Hagenia* woodland, occasionally in moorland, but always in damp sites, along streams; (1350–)1800–3150 m
CONSERVATION NOTES. Widespread; least concern (LC)

SYN. *A. cicutarium* sensu Hieron. in P.O.A. C: 83 (1895), *non* Sw.
   A. gracillimum Kuhn in Hochgebirgsfl. trop. Afr.: 193 (1892). Syntypes: Kenya, Laikipia, 1300–2000 m, *von Höhnel* 52 (B!, syn.) & Tanzania, Kilimanjaro, *Meyer* 276 (B!, syn.), *non* Col. (1890)
   A. kuhnianum C.Chr. in Z.A.E.: 16 (1910). Type: as for *A. gracillimum*, **syn. nov.**
   A. tenuifolium sensu Peter, F.D.O.-A.: 80 (1929), *non* D.Don

NOTE. *A. kuhnianum* falls well within the variability of this taxon, and I have made it into a synonym.
   A specimen from Tanzania, Mt Meru, Dec. 1966, *Richards* 21572! is very similar but more dissected: it is 4-pinnatisect. The lamina is 53–65 × 14–28 cm, pinnae are up to 14 × 7 cm, and the indusium is 0.3–0.4 mm wide. It came from moist forest at 2220 m, on the bank of a small stream.

63. **Asplenium pumilum** *Sw.*, Nov. Gen. & Sp. Pl.: 129 (1788). Type: Jamaica, without data, *Swartz* s.n. (S, holo.; UPS, iso.)

Terrestrial (in our area); rhizome short, erect, 3–5 mm diameter, with subulate clathrate lacerate-serrate scales 3–4.5 mm long, with a long hair-tip, usually with a black central stripe and margin of pale or translucent cells. Fronds tufted, erect, softly herbaceous, not proliferous. Stipe greenish above becoming dark brown below, 2–10 cm long, with scattered pale multicellular hairs. Lamina ovate to triangular in outline, 3–11 × 2–12 cm, 2–pinnatisect to 3-pinnatifid, the basal pinnae the largest, the apical pinna similar to the other ones. Pinnae membranous, rhombic to narrowly triangular, up to 6 × 3 cm, deeply pinnatifid into ovate to obcuneate, deeply or shallowly crenate obtuse lobes; with sparse transparent or whitish hair-like scales along the veins of both surfaces. Rachis greyish-green with prominent pale green wings and with sparse pale multicellular hairs. Sori 2–10 per lobe, set at a slight angle to the costa, linear, 2–5 mm long, straight or slightly curved; indusium translucent, 0.3–0.6 mm wide, slightly erose. Fig. 10: 7, p. 62.

subsp. **hymenophylloides** (*Fée*) *Schelpe* in Bol. Soc Brot., Sér. 2, 41: 210 (1967); Schelpe, F.Z. Pteridophyta: 178 (1970); Burrows, S. Afr. Ferns: 255, map, figs. (1990). Type: Ethiopia, Simien, Amba Sea, 2000 m, *Schimper* s.n. (B!, holo.; K!, iso.)

Frond thin in texture, light green in colour.

TANZANIA. Kigoma District: Mt Livandabe, May 1997, *Bidgood et al.* 4189!; Kilosa District: Mikumi National Park, Vuma Hills forest, June 1977, *Wingfield & Mhoro* 4019!; Mbeya District: Mbilizi–Galula road km 8, June 1996, *Faden et al.* 96/471!
DISTR. **T** 4, 6–8; W to NE tropical Africa and south to Zimbabwe; NW India
HAB. Banks of rivers and seasonal streams; 500–1400 m
CONSERVATION NOTES. Widespread; least concern (LC); uncommon in our area with five specimens seen from Tanzania.

SYN. *A. pumilum* Sw. var. *hymenophylloides* Fée, Mém. Fam. Foug. 7: 54, t. 15.4 (1857)
    A. eylesii Sim, Ferns S. Afr. ed. 2 : 147, t. 61 fig. 2 (1915). Type: Zimbabwe, Mazoe Distr., Iron Mask Hills, *Eyles* 564 (PRE, holotype; SRGH, isotype)

NOTE. The other subspecies, subsp. *pumilum*, occurs in the Americas. It has a thicker leaf texture and the lamina is dark green; the pinna margin is slightly less cut. Plants from the Americas are generally larger (see Pic. Serm. in Webbia 37 (1) (1983)).

64. **Asplenium stenopteron** *Peter*, F.D.O.-A. Descr.: 6, t. 2.3 (1929). Type: Tanzania, Kilimanjaro, Marangu to Bismarck Hut, *Peter* 598 (B!, holo.)

Terrestrial; rhizome erect, short with rather dense dark red-brown narrowly triangular attenuate scales to 4.5 × 0.7 mm. Fronds suberect, herbaceous, lanceolate, 19 × 4 cm, not proliferous on the single fertile frond seen. Stipe green, 5.5 cm long, with hair-shaped scales to 3 mm long, with an occasional lobe. Lamina to 13 cm long, 1-pinnate, pinnae in 10–11 pairs, alternate, lanceolate, the median largest and to 2.1 × 0.5 cm, the basal reduced, the apical gradually decrescent, the terminal pinna in line with the rachis, a lobed unit 2.7 × 0.8 mm, basiscopic base cuneate and almost dimidiate, acroscopic base at more of an angle, margin whitish, slightly thickened, crenate-serrate, apex narrowly obtuse; glabrous or with few minute scales. Rachis narrowly winged, with few scales as on stipe. Sori 1(–2) per pinna, ovate, 2.5–3 mm long; indusium entire, ± 0.5 mm wide.

TANZANIA. Kilimanjaro, Marangu to Bismarck Hut, *Peter* 598!
DISTR. **T** 2; only known from the type
HAB. No data; 1700 m
CONSERVATION NOTES. Possibly extinct; this is a well-collected area. A specific search is needed for the 'Ex' category.

NOTE. A taxon quite on its own, does not key anywhere close.

UNRESOLVED TAXA

Asplenium chlaenopteron Fée, Gen.: 194 (1850/52); Z.A.E.: 8 (1910)

Brause & Hieron. state this taxon occurs in Kilimanjaro and Bukoba. It is supposed to be really close to *A. boltonii*. The type is from Rwanda. I have seen no East African specimens.

Asplenium sp. F of Johns, Pterid. trop. East Africa checklist: 68 (1991); *Pócs* 6006B

Asplenium sp. G of Johns, Pterid. trop. East Africa checklist: 68 (1991); *Faden* 69/1112

Note: For neither of these any information or diagnosis is given by Johns.

Asplenium lademannianum *Rosenst.* in F.R. 6: 177 (1908); Pic.Serm. in B.J.B.B. 55: 138 (1985). Type: Tanzania, "Kondoa-Irangi, Mt Ufiome", *Lademann* s.n. (B/P?, holo. – not found) [Note Mt Ufiome = Mt Kwaraha, in **T** 2 Mbulu District)

Protologue: "sect. *Euasplenium*; by lamina dissection very similar to plate 225 in Beddome's Ferns of S.I. '*laserpitiifolium*' but really *A. nitidum.* Distinct from that by narrower, more linear and more acuminate lamina, longer-attenuate pinnae, almost membranous texture, and dense indument on axes with flaccid curly ramenta."

Description from Pichi Sermoilli and Chaerle: "rhizome erect, subwoody, 10–15 mm diameter, with flaccid membranous lanceolate entire acuminate rhizome scales ± 4 mm long, with a large hyaline lumen and dark brown narrow margins, ending in a hair-tip. Fronds to 58 × 16 cm; stipe brown or grey-green, 17–22 cm long, with reflexed scales similar to rhizome scales mixed with linear ones, glabrescent; lamina lanceolate in outline, membranous-herbaceous, 2-pinnate, the basal pair remote and slightly reduced, apical pinna reduced, linear, incised-lobed; pinnae in ± 20 pairs, alternate or subopposite, shortly petiolulate, narrowly ovate, to 8 × 2.5 cm, pinnate, caudate-acuminate, the pinnules in up to 8 pairs, to 2 × 1 cm, deeply incised-lobed, glabrous except for the scaly costae and veins; rachis brown, ± densely scaly with scales similar to stipe. Sori 2–4 per pinnule, 'short'; indusium membranous, entire."

TANZANIA. Mbulu District: Mt Ufiome, *Lademann* s.n.
DISTR. **T** 2/5; not known elsewhere
HAB. 'in silva primaeva', 1800 m

NOTE. Treated by Pichi Sermolli as full species; he saw part of the type – he also said most of the type was now missing at P ("the holotype was held in Rosenstock's herbarium and now ought to be in Paris herbarium but... (...) in addition to this I have found a pencil rubbing of two fronds in B, two pinnae and some rhizome scales"). He says the rhizome is not erect but shortly creeping with shortly spaced fronds; this is based on *Lambinon* 74/1540, which "agrees very well with the type". He adds that it resembles *A. aethiopicum* but is distinct in short-creeping horizontal rhizome, membranous and long-linear to narrowly linear-lanceolate rhizome scales; that small forms of *lademannianum* look like *A. lividum* (but differ by rhizome scales and gradual shapes). Larger *lademannianum* look like *A. volkensii* but differ in frond dissection and short-creeping, not long-creeping; plus rhizome scale characters. Also treated by Chaerle in his unpublished thesis as a full species, keying out as rhizome creeping, 15 mm thick, 11–22 pinnae pairs, as opposed to rhizome erect/short-creeping, thinner, 6–12 pinnae pairs for *A. aethiopicum.*

I am not convinced by all this. The part of the type with frond tracings at B could very well be *A. aethiopicum*; there is no indication at all whether the rhizome is erect or creeping. To equate specimens found far away I find very risky; I prefer to treat this taxon as unresolved.

Asplenium mettenii Kuhn, Fil. Afr.: 20, 107 (1968). Type: Kenya, near Mombasa, Jan. 1865, *Kersten* 29 (B!, holo.)

The type specimen lacks a rhizome and probably the entire stipe; it also lacks the frond apex. It consists of what I believe to be part of the rachis with 3? pinnae on one side, and 2 pinnae on the other. In an envelope there is one more pinna, plus what might be the frond apex (gradually decrescent, not proliferous). The pinnae are ± 3 cm apart, 2-pinnate (making the frond 3-pinnate) and up to 9.5 × 3.9 cm with up to 10 pinnules on each side, the lowermost with 2–3 segments on each side and up to 2.3 × 1.2 cm; the segments are serrate, with acute apices. Rachis unwinged, with sparse filiform scales. Sori 1 per serrate tooth, several per pinna, linear and 2–4 mm long; indusium dark translucent, membranous, to 0.4 mm wide, ± entire.

KENYA. Kenya, near Mombasa, Jan. 1865, *Kersten* 29!
DISTR. **K** 7; see Note
HAB. no data

NOTE. I have compared this to all taxa that seem to come close, and it does not really match anything. It seems a distinct taxon, but due to all the missing parts it will have to remain vague. Two specimens from the Comoro Islands are a good match: *Humblot* 210 (B!) and possibly *Humblot* 318 (B!), but again these lack all basal parts and the apex.

Asplenium ruwenzoriense *Baker* in K.B. 1901: 137 (1901); Johns, Pterid. trop. East Africa checklist: 67 (1991). Type: Uganda, Ruwenzori, Huata, *Scott-Elliot* 7706 (K!, holo.)

Basal parts not preserved. Lamina oblong-lanceolate, 60–90 cm long, 1-pinnate with at least 25 pairs of pinnae, the lower alternate, the upper opposite, the terminal ones gradually decrescent, green, glabrous. Pinnae linear, 10–13 × 1.2–1.4 cm, sessile or very shortly petiolate, base unequally cuneate and some with slight lobe on the acroscopic side, margin crenate, apex acuminate; veins ascending, forked. Rachis dark green, glabrous but for the occasional capillary scale. Sori linear-oblong, 4–5 mm long, not reaching the margin, restricted to lower part of vein; indusium to 1 mm wide, firm, entire, persistent.

UGANDA. Ruwenzori, Kivata, 1893-94, *Scott-Elliot* 7706!
DISTR. **U** 2; only known from the type
HAB. 'common in forests, 6000–8000 feet'

NOTE. As the type specimen lacks a rhizome, the stipe and the lower pinnae, there is not much to go on! It does not seem to be proliferous (though part of the apex is carefully glued behind a middle pinna) but if it was it might be *A. elliottii*; otherwise it keys out to *A. smedsii*, but the margin is not really incise enough for that species. Until more material is collected this will have to remain a mystery.

Asplenium sertularioides *Baker* in Trans. Linn. Soc. 2: 354 (1887); Johns, Pterid. trop. East Africa checklist: 67 (1991). Type: Tanzania, Kilimanjaro, 9000–13000', *Johnston* 26 (K!, holo.)

Growth form unknown; fronds tufted, other details unknown; rhizome scales (assumed from scales on lower stipe!) narrowly triangular, to 8 × 1 mm, attenuate at apex, or (assumed from very young *Grimshaw* frond) pale brown, ovate, to 8 × 2 mm, acute. Stipe up to at least 27 cm, with dark brown lobed scales to 8 mm. Lamina sub-coriaceous, oblong-lanceolate, 20–56 × 5–10 cm, 2-pinnatifid to 2–pinnatipartite to 3-pinnatifid with the acroscopic basal lobe often enlarged and almost free, lower pinnae reduced and only shallowly lobed. Pinnae in 24–35 pairs, lanceolate, 3.7–7.5 cm long, to 1 cm wide, pinnatifid, sessile, base unequal, acroscopic base with up to 8-lobed pinnule, costa shortly winged, lobes to 4 × 1.5 mm, glabous above, sparsely scaly beneath with narrow and several-thin-lobed scales to 1.5 mm long. Rachis with sparse scales to 4 mm long, with thin-lobed margin. Sori 1 per lobe, about halfway up the lobe or in lower part, on the costule and emergent from the acroscopic margin, 1–2 mm long; involucre 0.5–0.7 mm wide.

TANZANIA. Mt Meru, E slopes, Nov. 1948, *Hedberg* 2439!; Kilimanjaro, 9000–13000', 1884, *H.H. Johnston* 26! & between Horombo & Mandara Hut, Sep. 1993, *Grimshaw* 93.660b!
DISTR. **T** 2; not known elsewhere
HAB. Upper heath zone, mossy rock crevice in moorland; 3350–3500 m (excluded wide 'range' of type)

NOTE. Baker states in the protologue this taxon resembles most *A. belangeri* Kunze (which is an Asian species). I believe this is very much like *rutifolium* but the altitude is quite different, the stipe is too long, the pinnae too many; and for *loxoscaphoides* the pinnae are too short. Several specimens from high altitude in the Ruwenzori are superficially similar, but the sori are consistently long (3.5–4.5 mm). See next taxon, *Asplenium* 'D586'.

Pic. Serm. in B.J.B.B. 55: 151 (1985) treats specimens from the Virunga Mts, Rwanda as this taxon. He says it is allied to *A. loxoscaphoides* but differs mainly in broadly triangular reduced pinnae of the lower juga, and in the lowermost segments of the pinna, both the basiscopic and the acroscopic one, sharply distinct in shape and length from adjoining segments.

I prefer to treat this as 'unresolved' until complete material has been collected.

Asplenium 'D586'

Growth form unknown; rhizome details unknown. Not proliferous. Stipe 23–31+ cm, with mid or dark brown lobed scales to 6 × 2 mm near base. Lamina sub-coriaceous, oblong-lanceolate, 37–62 × 7–13 cm, 2-pinnatisect to 3-pinnatifid with the acroscopic basal lobe often enlarged, lower pinnae reduced and lobed more shallowly. Pinnae in 22–29 pairs, lanceolate, the largest 5–8.5 × 1–1.7 cm, pinnatifid, base unequal, acroscopic base with up to 5-lobed pinnule, costa shortly winged, lobes to 7 × 1.5 (–2) mm, glabous above, sparsely scaly beneath with narrow and several-thin-lobed scales to 1.5 mm long. Rachis with sparse scales to 2 mm long, with thin-lobed margin. Sori 1 per lobe, about halfway up the lobe or in lower part, on the costule and emergent from the acroscopic margin, 3–4.5 mm long; involucre to 0.8 mm wide.

Uganda. Ruwenzori, Oct. 1905, *Dawe* 586! & Mijusi valley, Mar. 1948, *Hedberg* 606! & Bujuku, Aug. 1931, *Hancock & Fishlock* 137! & Nyomleju, Jan. 1951, *Osmaston* 3661!
Distr. **U** 2; Ruwenzori only
Hab. Heath zone and open forest; (2700?–)3000–3500 m

Note. If creeping, keys to *aethiopicum* group, but not really belonging there. I prefer to treat this as 'unresolved' until complete material has been collected.

Asplenium 'L2775'

Lithophyte; rhizome erect, with scales ovate, to 3.5 × 1.4 mm. Fronds tufted, proliferous at apex. Stipe 21–29 cm long, with a few small scales near base. Lamina ovate, 20–30 × 16–17 cm, 1-pinnate, lower pinnae not reduced but largest. Pinnae in 4–5 pairs, lanceolate, the largest 8–10 × 1.7–1.9 cm, attenuate, margin serrate, glabous or nearly so; terminal pinna narrowly ovate, 12–18 × 3–6 cm, irregularly lobed and proliferous at the attenuate apex. Rachis with sparse scales to 1 mm long, with thin-lobed margin. Sori many per lobe, reaching neither costa nor margin, on the branched veins and (2.5–)5–10 mm long; involucre whitish, membranous, entire, to 0.6 mm wide.

Tanzania. Morogoro District: Kanga Mountain, Dec. 1987, *Lovett & Thomas* 2775!
Distr. **T** 6; not known elsewhere
Hab. On shaded rock in moist forest; 1200 m

Note. This slightly resembles *A. angolense* Baker but that has the gemmae in the centre of the terminal pinna; also, the dissection of the terminal pinna in this specimen is very different.

A specimen from **T** 2, Mt Meru, slopes of little Meru, Mar. 1967, *Vesey-Fitzgerald* 5130! comes from a damp shady ravine in forest shade, at altitude 2400 m. It has tufted fronds, is 4-pinnatifid, and non-proliferous. Sadly it is sterile, so I am unable to take this any further.

Schippers in Fern Gaz. 14, 6: 200 (1993) cites *Asplenium gautieri* Hook., based on Pócs & Nsolomo 87051/AG (not seen) – this taxon is otherwise only known from Madagascar.

EXCLUDED TAXA

Asplenium affine sensu Johns, Pterid. trop. East Africa checklist: 61 (1991), *non* Sw. in Schrad. Journ. 1800 [2]: 56 (1801)

This is a misnomer; the real *A. affine* is an Indo-Malesian species.

Asplenium caudatum sensu Johns, Pterid. trop. East Africa checklist: 62 (1991), *non* Forst.

The real *A. caudatum* is an Australian taxon. Many of our Kew specimens had been ID-d with this name, and most of these turned out to be *A. pellucidum* subsp. *pseudohorridum*.

Asplenium hemionitis *L.* var. **rotundatum** *Peter*, F.D.O.-A. Descr.: 5, t. 5.4 (1929). Type: "N Spanien oder Teneriffe", *Peter* 45234 (B!, holo.)

Not an East African species, though published in the Flora of German East Africa!

Asplenium hemitomum *Hieron.* in E.J. 46: 365 (1911); Tardieu in Mém. I.F.A.N. 28: 189 (1953); Alston, Ferns W.T.A.: 59 (1959); Tardieu, Fl. Cameroun Pterid.: 203, t. 31/5–8, t. 29/9 (1964); Johns, Pterid. trop. East Africa checklist: 63 (1991). Type: Bioko [Fernando Po], *Barter* s.n. (B!, lecto., K!, iso.), chosen by Tardieu?

Rhizome erect, short, with dark brown narrowly ovate attenuate rhizome scales. Fronds tufted, 20–50 cm long. Stipe dark grey, 10–23 cm long, canaliculate in upper part, bearing scales similar to those of rhizome. Lamina oblong, 1-pinnate, 10–35 × 7–14 cm. Pinnae in 5–12 pairs, at an angle of 85° with the rachis, (sub)opposite, sub-coriaceous, petiolate, spaced, rhomboid and asymmetrical, 5–8 × 2–3 cm, base unequal, the acroscopic base cuneate, subtruncate and ± auriculate, the basiscopic base oblong and rounded, sometimes also auriculate, margin crenate-dentate, median lobe elongate [?], deltoid and serrate; terminal pinna often 3-lobed with acute lobes; pinnae with small narrow scales on lower surface. Rachis compressed, green, sparsely scaly. Sori many, along the veins forming an angle of ± 30° with the costa, linear, 10–23 mm long, not reaching the margin or the costa, some opening towards the costa, some towards the margin; indusium membranous, entire, 0.5–0.7 mm wide.

DISTR. Guinea to Congo-Kinshasha. Some specimens had been named as this taxon, but turned out to be *A. warneckei*.

Asplenium hylophilum Hieron., P.O.A. C: 84 (1895). Type: Tanzania, Lushoto District: Usambara, Shagayu Forest near Mbaramu, *Holst* 2480 (B, holo., BM, K, iso.)

= Diplazium nemorale (*Baker*) *Schelpe* , see F.T.E.A. Woodsiaceae

"*Asplenium pseudoserra* Hieron." – Johns, Pterid. trop. East Africa checklist: 66 (1991).

Name not found; possibly a misprint for *pseudohorridum*?

Asplenium rukararense *Hieron.*, Z.A.E. 2: 12, t. 2/d–e (1910); Pic.Serm. in B.J.B.B. 55: 150 (1985); Johns, Pterid. trop. East Africa checklist: 67 (1991). Type: Rwanda, Rugege forest, source of Rukarara, *Mildbraed* 922 (B!, holo.)

Rhizome erect (type) or shortly creeping, to 3 mm diameter, with mid-brown membranous elongate-deltoid acute rhizome scales ending in hair-tip, to 8 × 2.5 mm, glabrescent. Leaves to 40 cm long, proliferous just below apex, the basal few pinnae slightly reduced, apex decrescent. Stipe grey, 10–20 cm long, sparsely scaly, the scales linear, to 3 × 0.5 mm. Lamina lanceolate in outline, 25–35 × 8–17 cm, 2-pinnatifid; pinnae in 15–20 pairs, to 9 × 2.4 cm, deeply pinnatifid, the base almost pinnate, linear-lanceolate, base subtruncate, the ultimate segments in 11–13 pairs, the basal ones bifurcate or trifid, the others emarginate, the lobes ovate with crenate-serrate apex, decurrent at base. Sori at the base of ultimate segments on the acroscopic side of the vein, 1.5–3 mm long; indusium membranous, entire, to 0.4 mm wide.

DISTR. Rwanda, Burundi

NOTE. Two Ugandan specimens formerly identified as this have been named by me as the closely related *A. preussii*, which has an erect rhizome.

Asplenium schweinfurthii *Baker* in Balfour, Bot. Socotra: 328, t. 100 (1888); Johns, Pterid. trop. East Africa checklist: 66 (1991)

Species from Soqotra; to be excluded from FTEA.

Asplenium simii *Braithwaite & Schelpe* in Bol. Soc Brot., Sér. 2, 41: 209 (1967); Schelpe, F.Z. Pteridophyta: 181 (1970); Burrows, S. Afr. Ferns: 250, map, figs. (1990); Johns, Pterid. trop. East Africa checklist: 67 (1991). Type: Rhodesia: Vumba Mts, Elephant Forest, *Chase* 6274 (BOL, holotype; SRGH, isotype)

DISTR. Mozambique, Zimbabwe, South Africa

Does not occur in East Africa; specimens identified as this taxon were *A. stuhlmannii* or others. I have not seen *Schippers* 1560 from **T** 3, South Pare Mts., which Schippers in Fern Gaz. 14, 6: 203 (1993) and Chaerle, in his unpublished thesis, p. 218, include under this taxon.

Asplenium thunbergii *Kunze* in Linnaea 10: 517 (1836); Johns, Pterid. trop. East Africa checklist: 67 (1991)

Synonym *Caenopteris auriculata* Thunb. in Nova Acta Petr. 9: 159, t. E. 2 (1795), *non* Asplenium auriculatum Sw. 1817; Type from South Africa, Cape of Good Hope. Not an East African taxon.

Asplenium adscensionis (Forst.) Bernh. = Pteris dentata subsp. flabellata (Thunb.) Runemark
Asplenium aquilinum (L.) Bernh.= Pteridium aquilinum (L.) Kuhn Key
Asplenium cordatum (Thunb.) Sw. = Ceterach cordatum (Thunb.) Desv.
Asplenium hyophilum Hieron. = Diplazium nemorale (Baker) Schelpe
Asplenium laxum (Pappe & Raws.) Kuhn = Athyrium scandicinum (Willd.) C.Presl
Asplenium leptophyllum (L.) Sw. = Anogramma leptophylla (L.) Link
Asplenium madagascariense Baker = Diplazium nemorale (Baker) Schelpe
Asplenium mary-annae Kunkel = Coniogramme africana Hieron.
Asplenium nemorale Baker = Diplazium nemorale (Baker) Schelpe
Asplenium polydactylon Webb = Actiniopteris radiata (Sw.) Link
Asplenium radiatum Sw. = Actiniopteris radiata (Sw.) Link
Asplenium scandicinum (Willd.) Heller = Athyrium scandicinum (Willd.) C.Presl
Asplenium schimperi (Fée) A. Braun = Athyrium schimperi Fée
Asplenium serrulatum (Sw.) Bernh. = Xiphopteris serrulata (Sw.) Kaulf.
Asplenium zanzibaricum Baker = Diplazium zanzibaricum (Baker) C.Chr.

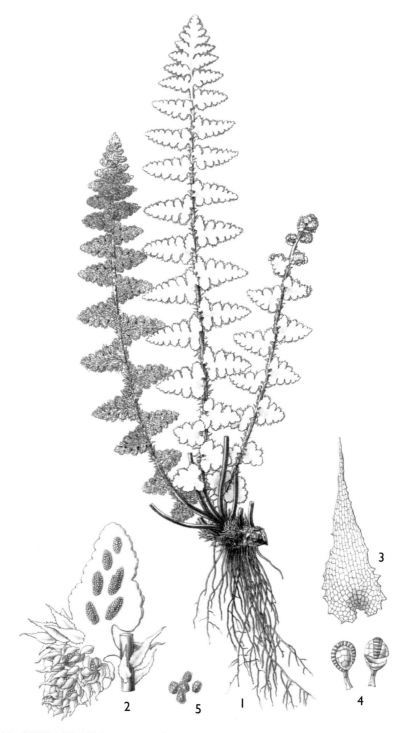

FIG. 11. *CETERACH CORDATUM* — **1**, habit × 1; **2**, Part pinna with sori × 6; **3**, scale of frond × 12; **4**, sporangia enlarged, not to scale; **5**, spores enlarged, not to scale. Reproduced from Hooker & Greville, Icones Filicum, tab. 156 (1829).

2. CETERACH

DC. in Lam. & DC., Fl. Fr., ed. 3, 2: 566 (1805)

Rhizome erect, short, with clathrate scales. Stipes short, tufted, densely scaly. Frond usually deeply pinnatifid to 2-pinnatifid, glabrous ventrally at maturity but densely scaly on the lower surface; veins usually anastomosing marginally. Sori elongate along the veins; indusium inconspicuous or obsolete.

A genus of ± 5 species in Europe, Asia and Africa.

Sometimes considered as a synonym of *Asplenium.*

Ceterach cordatum (*Thunb.*) *Desv.* in Mém. Soc Linn. Par. 6, 2: 223 (1827); Sim, Ferns S. Afr. ed. 2: 175, t. 73 (1915); Schelpe, F.Z. Pteridophyta: 188, t. 54f (1970); Burrows, S. Afr. Ferns: 255, map, figs. (1990); Thulin, Fl. Somal. 1: 14 (1993); Faden in U.K.W.F. ed. 2: 30 (1994). Type: South Africa, Cape Province, *Thunberg* s.n. (UPS, holo.)

Terrestrial; rhizome erect or procumbent, to 4 mm diameter, with dark brown clathrate lanceolate pseudoserrate rhizome scales to 4.5 × 1 mm, attenuate at apex. Fronds tufted, suberect, not proliferous, coriaceous, curling and inrolled when dry. Stipe redbrown, 4–20 mm long, with dense brown lanceolate acuminate scales to 2 mm long. Lamina bright to dark green, elliptic to narrowly elliptic, to 24 × 8.5 cm, 1-pinnate to 2-pinnatifid, lower pinnae gradually reduced, diminishing gradually to apex; glabrous on upper surface, densely set with pale brown ovate-acuminate to lanceolate scales to 3 × 1.5 mm on lower surface; pinnae 7–11(–15) pairs, oblong, to 2.5 × 0.9 cm, base adnate to free and auriculate, weakly undulate to pinnatifid with broadly oblong crenate segments, apex rounded; rachis covered in brown scales. Sori several per pinna on each side of the costa but almost completely hidden by scales, linear, to 2 mm long. Fig. 11, p. 70.

UGANDA. Karamoja District: Kangole–Nabilatuk, July 1949, *H.D. van Someren* 539!
KENYA. Samburu District: Kirimun escarpment 60 km N of Rumuruti, Apr. 1975, *Hepper & Field* 5096!; Naivasha District: Gilgil–Elmenteita road, Oct. 1972, *Gillett* 10061!; Machakos District: Lukenya, May 1981, *Gilbert* 6132!
TANZANIA. Musoma District: Mara River guard post, Dec. 1964, *Greenway & Turner* 11780! & 24 km above Mara River guard post, Oct. 1961, *Greenway & Turner* 10242!
DISTR. **U** 1; **K** 1–4, 6; **T** 1, 3 (see note); tropical Africa from Ethiopia and Somalia to South Africa
HAB. Rock crevices and around bases of boulders, riverine thicket, always in sites that dry out for long periods; may be locally common; 1000–1950 m
CONSERVATION NOTES. Widespread; least concern (LC)

SYN. *Acrostichum cordatum* Thunb., Prodr. Pl. Cap.: 171 (1800)
    Asplenium cordatum (Thunb.) Sw. in Schrad., Journ. Bot. 1800, 2: 54 (1801)
    Grammitis cordata (Thunb.) Sw., Syn. Fil.: 23, 217 (1806)
    Cincinalis cordata (Thunb.) Desv. in Mag. Ges. Naturf. Fr. Berl. 5: 311 (1811)
    Notholaena cordata (Thunb.) Desv. in Journ. de Bot. 1, App.: 92 (1813)
    Gymnogramma cordata (Thunb.) Schlechtend., Adumbr.: 16 (1825)
    Ceterach phillipsianum Kümmerle in Botan. Kozl. 6: 287 (1909) & Mag. Bot. Lap. 8: 354 (1909). Type: Somalia, *Lort Phillips* s.n. (K, iso.)

NOTE. Schippers in Fern Gaz. 14, 6: 202 (1993) states he has seen this on a rock wall in North Pare, which would add **T** 3 to the distribution.
    There have been proposals to split this taxon into several entities, but I have yet to see a key to such entities that works. Until that time, I will treat this as a single variable species.

INDEX TO ASPLENIACEAE

Lonchitis bipinnata Forssk., 56
Loxoscaphe concinnum (Schrad.) Moore, 55
Loxoscaphe Moore, 2
Loxoscaphe nigrescens (Hook.) Moore, 49
Loxoscaphe spathulata Pic.Serm., 49
Loxoscaphe theciferum (H.B.K.) T.Moore
  var. *concinna* (Schrad.) C.Chr., 55

Notholaena cordata (Thunb.) Desv., 71

Tarachia friesiorum (C.Chr.) Momose, 28
Tarachia furcata (Thunb.) C.Presl, 59
Trichomanes aethiopicum Burm. f., 59

New names validated in this part

Asplenium udzungwense *Beentje*, **sp. nov.**

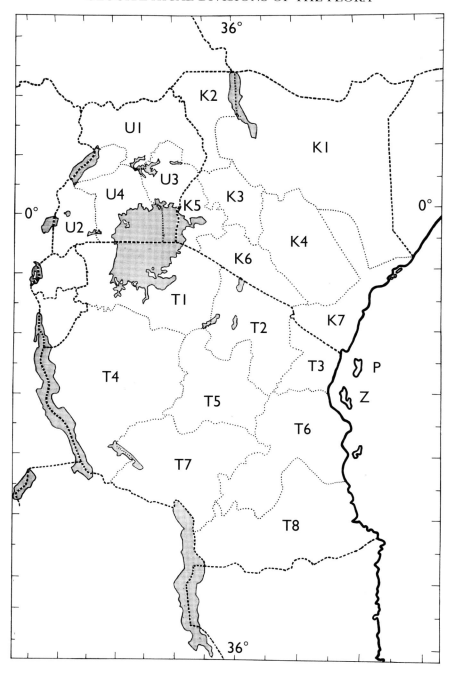

PLANTS PEOPLE
POSSIBILITIES

First published in 2008 by
Royal Botanic Gardens, Kew
Richmond, Surrey, TW9 3AB, UK
www.kew.org

ISBN 978 1 84246 202 7

British Library Cataloguing in Publication Data
A catalogue record for this book is available from the British Library

Design and typesetting by Margaret Newman,
Kew Publishing, Royal Botanic Gardens, Kew.

For information or to purchase all Kew titles please visit
www.kewbooks.com or email publishing@kew.org

All proceeds go to support Kew's work in saving the world's plants for life

LIST OF ABBREVIATIONS

A.V.P. = O. Hedberg, Afroalpine Vascular Plants; **B.J.B.B.** = Bulletin du Jardin Botanique de l'Etat, Bruxelles; Bulletin du Jardin Botanique Nationale de Belgique; **B.S.B.B.** = Bulletin de la Société Royale de Botanique de Belgique; **C.F.A.** = Conspectus Florae Angolensis; **E.J.** = A. Engler, Botanische Jahrbücher für Systematik, Pflanzengeschichte und Pflanzengeographie; **E.M.** = A. Engler, Monographieen Afrikanischer Pflanzen-Familien und Gattungen; **E.P.** = A. Engler, Das Pflanzenreich; **E.P.A.** = G. Cufodontis, Enumeratio Plantarum Aethiopiae Spermatophyta; in B.J.B.B. 23, Suppl. (1953) et seq.; **E. & P. Pf.** = A. Engler & K. Prantl, Die Natürlichen Pflanzenfamilien; **F.A.C.** = Flore d'Afrique Centrale (*formerly* F.C.B.); **F.C.B.** = Flore du Congo Belge et du Ruanda-Urundi; Flore du Congo, du Rwanda et du Burundi; **F.E.E.** = Flora of Ethiopia & Eritrea; **F.D.-O.A.** = A. Peter, Flora von Deutsch-Ostafrika; **F.F.N.R.** = F. White, Forest Flora of Northern Rhodesia; **F.P.N.A.** = W. Robyns, Flore des Spermatophytes du Parc National Albert; **F.P.S.** = F.W. Andrews, Flowering Plants of the Anglo-Egyptian Sudan *or* Flowering Plants of the Sudan; **F.P.U.** = E. Lind & A. Tallantire, Some Common Flowering Plants of Uganda; **F.R.** = F. Fedde, Repertorium Speciorum Novarum Regni Vegetabilis; **F.S.A.** = Flora of Southern Africa; **F.T.A.** = Flora of Tropical Africa; **F.W.T.A.** = Flora of West Tropical Africa; **F.Z.** = Flora Zambesiaca; **G.F.P.** = J. Hutchinson, The Genera of Flowering Plants; **G.P.** = G. Bentham & J.D. Hooker, Genera Plantarum; **G.T.** = D.M. Napper, Grasses of Tanganyika; **I.G.U.** = K.W. Harker & D.M. Napper, An Illustrated Guide to the Grasses of Uganda; **I.T.U.** = W.J. Eggeling, Indigenous Trees of the Uganda Protectorate; **J.B.** = Journal of Botany; **J.L.S.** = Journal of the Linnean Society of London, Botany; **K.B.** = Kew Bulletin, *or* Bulletin of Miscellaneous Information, Kew; **K.T.S.** = I. Dale & P.J. Greenway, Kenya Trees and Shrubs; **K.T.S.L.** = H.J. Beentje, Kenya Trees, Shrubs and Lianas; **L.T.A.** = E.G. Baker, Leguminosae of Tropical Africa; **N.B.G.B.** = Notizblatt des Botanischen Gartens und Museums zu Berlin-Dahlem; **P.O.A.** = A. Engler, Die Pflanzenwelt Ost-Afrikas und der Nachbargebiete; **R.K.G.** = A.V. Bogdan, A Revised List of Kenya Grasses; **T.S.K.** = E. Battiscombe, Trees and Shrubs of Kenya Colony; **T.T.C.L.** = J.P.M. Brenan, Check-lists of the Forest Trees and Shrubs of the British Empire no. 5, part II, Tanganyika Territory; **U.K.W.F.** = A.D.Q. Agnew (or for ed. 2, A.D.Q. Agnew & S. Agnew), Upland Kenya Wild Flowers; **U.O.P.Z.** = R.O. Williams, Useful and Ornamental Plants in Zanzibar and Pemba; **V.E.** = A. Engler & O. Drude, Die Vegetation der Erde, IX, Pflanzenwelt Afrikas; **W.F.K.** = A.J. Jex-Blake, Some Wild Flowers of Kenya; **Z.A.E.** = Wissenschaftliche Ergebnisse der Deutschen Zentral-Afrika-Expedition 1907–1908, 2 (Botanik).

FAMILIES OF VASCULAR PLANTS REPRESENTED IN
THE FLORA OF TROPICAL EAST AFRICA

The family system used in the Flora has diverged in some respects from that now in use at Kew and the herbaria in East Africa. The accepted family name of a synonym or alternative is indicated by the word "see". Included family names are referred to the one used in the Flora by "in" if in accordance with the current system, and "as" if not. Where two families are included in one fascicle the subsidiary family is referred to the main family by "with".

PUBLISHED PARTS

Foreword and preface
*Glossary
Index of Collecting Localities

Acanthaceae
  Part 1
*Actiniopteridaceae
*Adiantaceae
Aizoaceae
Alangiaceae
Alismataceae
*Alliaceae
*Aloaceae
*Amaranthaceae
*Amaryllidaceae
*Anacardiaceae
*Ancistrocladaceae
Anisophyllaceae — as Rhizophoraceae
Annonaceae
*Anthericaceae
Apiaceae — see Umbelliferae
Apocynaceae
  *Part 1
*Aponogetonaceae
*Aquifoliaceae
*Araceae
Araliaceae
Arecaceae — see Palmae
*Aristolochiaceae
Asparagaceae
*Asphodelaceae
Aspleniaceae
Asteraceae — see Compositae
Avicenniaceae — as Verbenaceae
*Azollaceae

*Balanitaceae
*Balanophoraceae

*Balsaminaceae
Basellaceae
Begoniaceae
Berberidaceae
Bignoniaceae
Bischofiaceae — in Euphorbiaceae
Bixaceae
Blechnaceae
*Bombacaceae
*Boraginaceae
Brassicaceae — see Cruciferae
Brexiaceae
Buddlejaceae — as Loganiaceae
*Burmanniaceae
*Burseraceae
Butomaceae
Buxaceae

Cabombaceae
Cactaceae
Caesalpiniaceae — in Leguminosae
*Callitrichaceae
Campanulaceae
Canellaceae
Cannabaceae
Cannaceae — with Musaceae
Capparaceae
Caprifoliaceae
Caricaceae
Caryophyllaceae
*Casuarinaceae
Cecropiaceae — with Moraceae
*Celastraceae
*Ceratophyllaceae
Chenopodiaceae
Chrysobalanaceae — as Rosaceae
Clusiaceae — see Guttiferae
Cobaeaceae — with Bignoniaceae
Cochlospermaceae

Papaveraceae
Papilionaceae — in Leguminosae
*Parkeriaceae
Passifloraceae
Pedaliaceae
Periplocaceae — see Apocynaceae (Part 2)
Phytolaccaceae
*Piperaceae
Pittosporaceae
Plantaginaceae
Plumbaginaceae
Poaceae — see Gramineae
Podocarpaceae
Podostemaceae
Polemoniaceae — see Cobaeaceae
Polygalaceae
Polygonaceae
*Polypodiaceae
Pontederiaceae
*Portulacaceae
Potamogetonaceae
Primulaceae
*Proteaceae
*Psilotaceae
*Ptaeroxylaceae
*Pteridaceae

*Rafflesiaceae
Ranunculaceae
Resedaceae
Restionaceae
Rhamnaceae
Rhizophoraceae
Rosaceae
Rubiaceae
    Part 1
    *Part 2
    *Part 3
*Ruppiaceae
*Rutaceae

*Salicaceae
Salvadoraceae
*Salviniaceae
Santalaceae
*Sapindaceae
Sapotaceae
*Schizaeaceae
Scrophulariaceae

Scytopetalaceae
Selaginellaceae
Selaginaceae — in Scrophulariaceae
*Simaroubaceae
*Smilacaceae
Sonneratiaceae
Sphenocleaceae
Strychnaceae — in Loganiaceae
*Surianaceae
Sterculiaceae

Taccaceae
Tamaricaceae
Tecophilaeaceae
Ternstroemiaceae — in Theaceae
Tetragoniaceae — in Aizoaceae
Theaceae
Thelypteridaceae
Thismiaceae — in Burmanniaceae
Thymelaeaceae
*Tiliaceae
Trapaceae
Tribulaceae — in Zygophyllaceae
*Triuridaceae
Turneraceae
Typhaceae

Uapacaceae — in Euphorbiaceae
Ulmaceae
*Umbelliferae
*Urticaceae

Vacciniaceae — in Ericaceae
Valerianaceae
Velloziaceae
*Verbenaceae
*Violaceae
*Viscaceae
*Vitaceae
*Vittariaceae

*Woodsiaceae

*Xyridaceae

*Zannichelliaceae
*Zingiberaceae
*Zosteraceae
*Zygophyllaceae

FORTHCOMING PARTS

Acanthaceae
    Part 2
Apocynaceae
    Part 2

Asclepiadaceae — see Apocynaceae
Commelinaceae
Cyperaceae
Solanaceae

Editorial adviser, National Museums of Kenya: Quentin Luke
Adviser on Linnaean types: C. Jarvis

Parts of this Flora, unless otherwise indicated, are obtainable from:
Royal Botanic Gardens, Kew, Richmond, Surrey TW9 3AB, England. www.kew.org or www.kewbooks.com

* only available through CRC Press at:
UK and Rest of World (except North and South America):
CRS Press/ITPS,
Cheriton House, North Way, Andover, Hants SP10 5BE.
e: uk.tandf@thomsonpublishingservices. co.uk

North and South America:
CRC Press,
2000NW Corporate Blvd, Boco Raton, FL 33431-9868,
USA.
e: orders@crcpress.com

Information on current prices can be found at www.kewbooks.com or www.tandf.co.uk/books/